3rd International Symposium on Laser Ultrasonics and Advanced Sensing 2013

Journal of Physics: Conference Series Volume 520

Yokohama, Japan
25-28 June 2013

ISBN: 978-1-63439-301-0
ISSN: 1742-6588

3rd International Symposium on Laser Ultrasonics and Advanced Sensing 2013

Journal of Physics: Conference Series Volume 520

Yokohama, Japan
25-28 June 2013

ISBN: 978-1-63439-301-0
ISSN: 1742-6588

IOP Institute of Physics

Printed from e-media with permission by:

Curran Associates, Inc.
57 Morehouse Lane
Red Hook, NY 12571

Some format issues inherent in the e-media version may also appear in this print version.

Printed by Curran Associates, Inc. (2014)

For permission requests, please contact the Institute of Physics
at the address below.

Institute of Physics
Dirac House, Temple Back
Bristol BS1 6BE UK

Phone: 44 1 17 929 7481
Fax: 44 1 17 920 0979

techtracking@iop.org

Additional copies of this publication are available from:

Curran Associates, Inc.
57 Morehouse Lane
Red Hook, NY 12571 USA
Phone: 845-758-0400
Fax: 845-758-2634
Email: curran@proceedings.com
Web: www.proceedings.com

TABLE OF CONTENTS

PAPERS

LASER GENERATION AND/OR DETECTION

ULTRAFAST OPTOACOUSTICS

FLAW INSPECTION AND NONDESTRUCTIVE TESTING

MATERIAL EVALUATION AND CHARACTERIZATION

SIGNAL PROCESSING, MODELING AND IMAGING

ADVANCED SENSING INCLUDING NONCONTACT, MICRO / NANOSCALE OR NONLINEAR MEASUREMENTS

GUIDED WAVES

BIOMEDICAL APPLICATIONS

Preface

Based on the use of laser as a coherent and intense light source, the photo-acoustics originated from the discovery made by Alexander Graham Bell was extended to laser-ultrasonics (LU), and it has been applied to wide area of ultrasonics, optics, material characterization and nondestructive inspection.

In 1996, a research group for LU was started in the Japanese Society for Nondestructive Inspection (JSNDI), and researches on LU and related topics such as noncontact measurements and elastic wave theories were discussed. Similar activities were pursued also in North America and in Europe. The international symposium on LU was started in Montreal, Canada in 2008 by Jean Pierre Monchalin in order to offer a forum for involved with basic researches and industrial applications of LU. In the second symposium in Bordeaux, France nearly 120 papers were presented. It is our honor to have organized the third symposium, LU2013 on 25-28 June in Yokohama, Japan. The articles published here provide a sample of achievements presented there.

In LU2013, we focused on the laser generation and/or detection of acoustic waves, application to nondestructive testing, ultrafast-optoacoustics and innovative instruments. Research achievements in biomedical applications, advanced sensing including noncontact, micro/nanoscale or nonlinear measurements, as well as theory and simulation of ultrasound were also included, considering the interdisciplinary nature of this field.

We enjoyed very excellent and informative 3 plenary talks, 11 invited talks, 81 oral and 41 poster presentations with 168 attendees. According to requests, we organized a post deadline poster session to give an opportunity to present recent achievements after the deadline. Contributions of the participants, the scientific and organizing committees are highly appreciated. The conference tour was a dinner cruise to the Tokyo bay, and we hope this experience will remain as a pleasant memory in attendees. As decided in the meeting for the next symposium, the next symposium will be held in Chicago, USA in 2015. We would be happy if the articles in this issue work as a trigger to attend LU2015.

Kazushi Yamanaka
Ikuo Ihara
Makoto Ochiai

Organizing Committee

Chair	Kazushi Yamanaka	Tohoku University
	Ikuo Ihara	Nagaoka University of Technology
	Hirotsugu Inoue	Tokyo Institute of Technology
	Hideharu Minegishi	The Japanese Society for Non-Destructive Inspection
	Makoto Ochiai	Toshiba

Steering Committee

Chair	Makoto Ochiai	Toshiba
	Hideo Cho	Aoyama Gakuin University
	Ikuo Ihara	Nagaoka University of Technology
	Youichi Matsuda	Advanced Industrial Science and Technology
	Osamu Matsuda	Hokkaido University
	Hideo Nishino	The University of Tokushima
	Jun Semboshi	Toshiba
	Hisashi Yamawaki	National Institute for Materials Science
Secretary	Toshihiko Ohnishi	The Japanese Society for Non-Destructive Inspection
	Shohei Ooka	The Japanese Society for Non-Destructive Inspection

Organizer
The Japanese Society for Non-Destructive Inspection (JSNDI)
http://www.jsndi.jp/e/index.html

Scientific Committee

Chair	Ikuo Ihara	Nagaoka University of Technology, Japan
	Bertrand Audoin	University Bordeaux 1, France
	Paul C. Beard	University College London, UK
	Shiro Biwa	Kyoto University, Japan
	Alain Blouin	National Research Council of Canada, Canada
	Claude Boccara	City of Paris Industrial Physics and Chemistry Higher Educational Institution, France
	Emmanuel Bossy	City of Paris Industrial Physics and Chemistry Higher Educational Institution, France
	Hideo Cho	Aoyama Gakuin University, Japan
	Steve Dixon	University of Warwick, UK
	Manabu Enoki	The University of Tokyo, Japan

Journal of Physics: Conference Series **520** (2014) 011001 doi:10.1088/1742-6596/520/1/011001

Christ Glorieux	Catholic University of Leuven, Belgium
Vitalyi Gusev	University of Maine, France
Takahiro Hayashi	Kyoto University, Japan
Peter Hess	Heidelberg University, Germany
Souichi Hirose	Tokyo Institute of Technology, Japan
David Hutchins	University of Warwick, UK
Alexander A. Karabutov	Moscow State University, Russia
Sridhar Krishnaswamy	Northwestern University, USA
Seung-Seok Lee	Korea Research Institute of Standards and Science, Korea
Osamu Matsuda	Hokkaido University, Japan
Youichi Matsuda	Advanced Industrial Science and Technology, Japan
Tsuyoshi Mihara	University of Toyama, Japan
Jean-Pierre Monchalin	National Research Council of Canada, Canada
Todd Murray	University of Colorado, USA
Yasuaki Nagata	Nippon Steel and Sumitomo Metal, Japan
Kazuyuki Nakahata	Ehime University, Japan
Hideo Nishino	The Universitiy of Tokushima, Japan
Makoto Ochiai	Toshiba, Japan
Hirotsugu Ogi	Osaka University, Japan
Yukio Ogura	Japan Probe, Japan
Yoshikazu Ohara	Tohoku University, Japan
Yoshihumi Saijyo	Tohoku University, Japan
Jun Semboshi	Toshiba, Japan
Junji Takatsubo	Advanced Industrial Science and Technology, Japan
Motonobu Tomoda	Hokkaido University, Japan
Herve Tretout	Dassault Aviation, France
Shin-ichiro Umemura	Tohoku University, Japan
Hubert Voillaume	European Aeronautic Defence and Space, France
Shusou Wadaka	Ryoden Shonan Electronics, Japan
Oliver B. Wright	Hokkaido University, Japan
Kazushi Yamanaka	Tohoku University, Japan
Hisashi Yamawaki	National Institute for Materials Science, Japan
Shu-yi Zhang	Nanjing University, China

Journal of Physics: Conference Series **520** (2014) 011001 doi:10.1088/1742-6596/520/1/011001

Journal of Physics: Conference Series **520** (2014) 011002 doi:10.1088/1742-6596/520/1/011002

Peer review statement

All papers published in this volume of *Journal of Physics: Conference Series* have been peer reviewed through processes administered by the proceedings Editors. Reviews were conducted by expert referees to the professional and scientific standards expected of a proceedings journal published by IOP Publishing.

Journal of Physics: Conference Series **520** (2014) 012001 doi:10.1088/1742-6596/520/1/012001

Converging Laser Generated Ultrasonic Waves using Annular Patterns Irradiation

S Zamiri[1,2], B Reitinger[1,2], S Bauer[3] and P Burgholzer [1,2]

1-Christian Doppler Laboratory of Photoacoustic Imaging and Laser Ultrasonic, Science Park 2, Altenberger Straße 69, 4040 Linz, Austria
2-Research Center for Non Destructive Testing GmbH (RECENDT), Science Park 2, Altenberger Straße 69, 4040 Linz, Austria
3- Department of Soft Matter Physics, Johannes Kepler Universität, 4040 Linz, Austria
E-mail: saeid.zamiri@recendt.at

Abstract: We report on a contactless method to focus laser generated bulk and surface ultrasound waves in the thermo-elastic coupling mode by annular shaped illumination. By using a spatial light modulator (SLM) the beam profile of a pulsed picoseconds laser was shaped to annular forms flexibly and further rings with a thickness of 50 µm and a generation energy as low as 2 mJ were generated on the surface of aluminum plates. The annular shapes have been used to focus acoustic waves toward the center. In this work, a photorefractive adaptive interferometer set up based on Two-Wave Mixing in a fast BSO crystal was used to probe and detect the converging acoustic pulses at the center of the laser generated rings. By moving the detection point about 1 mm out of the ring epicenter, the amplitude of bulk and surface waves drop quickly which shows the converging evidence of the acoustic waves in the ring center. For a 3 mm thick aluminum plate, the ring size from 1 mm to 10 mm was scanned. The optimum ring diameter and the focal length of the acoustic waves along the central axis were found. Applications of this technique in subsurface defects detection as well as sample thickness measurement are investigated.

1. Introduction

Generation and detection of ultrasound pulses by using lasers is a valuable method for non-destructive testing of materials[1]. In this technique optically a pulsed laser with a short pulse length of picoseconds or even shorter is focused in a point on a material and due to local heating, ultrasonic waves with amplitude in order of nanometer are generated and propagated in all directions in the material[1,2]. By increasing the intensity of generation laser in one point on the sample surface, one can reach to the ablation mode. In this regime, a very thin layer of the surface is melted and the ultrasonic pulses with sufficient amplitude are generated in the material. For many industrial applications, detection of acoustic wave in the thermo-elastic generation region at very low incident energy and low amplitude is required which can be performed without destruction and melting the surface of sample.

One method to enhance the acoustic pulses amplitude at thermo-elastic mode is focusing the generated ultrasonic waves[3]. By changing the beam profile of a pulsed generation laser to an annular shape, ultrasonic pulses with spatial distributions are converged in the main axis of the annular shape [3]. This simple converging method by combining all generated signals into a point provides acoustic waves with efficient amplitude and furthermore ultrasonic measurements are performed with very low generation energy density. In this method because of spreading the short pulses energy in the beam shaped area, no ablation effect occurs in material [1] and the ultrasonic waves are generated in photo-thermal coupling regime.

C. K. Jen and coworkers reported on focused surface acoustic waves utilizing an axicon lens to shape the beam profile of a Q-switched YAG pulsed laser with energy of 100 mJ and 10 ns pulse width

Journal of Physics: Conference Series **520** (2014) 012001 doi:10.1088/1742-6596/520/1/012001

[2,3]. By changing the position of the axicon lens in respect to the sample surface, they generated annular forms on the surface a polished sample with diameter about 16 mm. They presented an amplification factor of 20 for surface waves detected in the center of a full ring generated in comparison to the line generation with the same laser fluency. A Michelson interferometer combined with a He-Ne laser was used to detect the generated surface waves in the center of the rings. X. Wang and coworkers used a mask to shaped the beam profile of a Q-switched Ti:Al_2O_3 pulsed laser with wavelength of 800 nm and pulse width of 50 ns [4]. They masked the laser output to a ring and by using a convex lens projected 0.5 mm thick rings on the surface of an aluminum plate. The rings radii were adjusted by changing the lens-sample distance. The generated bulk ultrasound waves were detected by contacting a piezoelectric transducer on the backside of the samples.

In the presented work here, we used a SLM which is a programmable electro optic device via a computer to flexibly shape and change the beam profile of a pulsed laser [5] to any arbitrary form such as point, line, elliptic as well as ring on the surface of an aluminum plate. By generation of rings with thickness of about 50 μm and programmable diameters, we were able to generate and focus the high frequency ultrasonic longitudinal pulses and surface acoustic waves.

2. Experiments and results

For detection of ultrasonic waves in the center of generated rings, a high sensitive photorefractive (PR) receiver set up with a BSO crystal based on two wave mixing method (TWM) in the crystal was used. In this ultrasound detector, a reference beam with a plane wave front and a reflected objective beam from the sample surface with a distorted wave front interfere in the BSO crystal [6]. Because of the PR effect, the refractive index of the crystal is modulated and due to generation a grating hologram in the PR material, two beams are coupled and diffracted. The wave front of the diffracted reference beam is adapted to the wave front of the transmitted signal beam. A fast photodiode is used to detect the modulated intensity of the diffracted and transmitted beams after the PR crystal. We applied this combined ring shaped illumination-PR detector set up for the sample thickness evaluation and subsurface defects inspection in a 3 mm aluminum plate.

A CW Nd: YAG Klastech laser with power of 300 mW and wavelength of 532 nm was used to detect the ultrasonic displacements on the sample surface. To increase the overlap distance between the reference and the signal beam inside the crystal, the reference beam by using a Galilean telescope was expanded to cover the BSO crystal. The polarizing beam splitters (PBS) and wave plates (HWP, QWP) were used to adjust the signal - reference beam intensity ratio. The spot size of the detection beam on the sample surface was about 200 μm. The interfered beams after the crystal are focused by means of a lens on the 125 MHz Photo-receiver.

In Fig.1, the schematic of the SLM experimental set up for ring generation combined with the PR adaptive interferometer for detection of ultrasound pulses in the ring center is shown. As a sample, a 3 mm thick aluminum plate with a rough surface was used. In Fig.2A, typical received ultrasonic waves in the center and close to the center of a generated ring with diameter of 3 mm and generation energy of 2 mJ/pulse are shown. As clearly is seen, when detection point is moved about 0.5 mm out of the center, two small surface waves are detected (dashed line). While at the ring center, a surface wave with high amplitude and first reflected longitudinal wave from the backside of the sample are received (solid line). In Fig.2B, peak to peak amplitude of the detected bulk and surface acoustic waves (inset figure) in the center of a 3 mm diameter ring as a function of distance from the center are displayed. As is seen, the amplitude of both detected bulk and surface wave drop quickly when the detection point is moved away from the ring center with a same manner. This shows that by use of the annular shape generation, the ultrasonic waves are focused at the center.

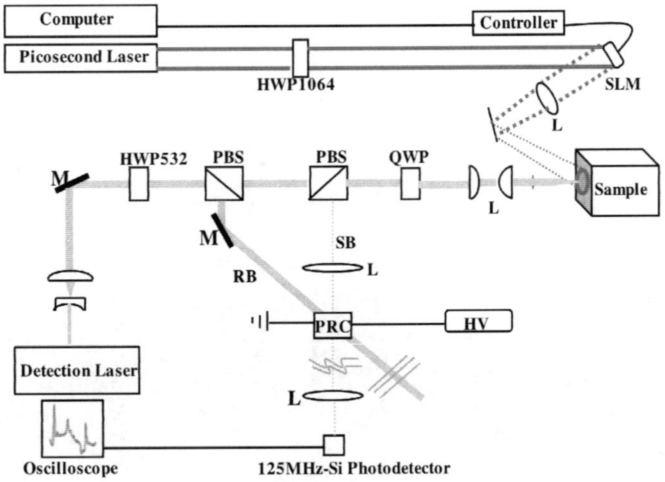

Figure 1: Simple schematic diagram of the experimental setup, HWPs: half wave plate for 532 nm and 1064 nm wavelength, QWP: quarter wave plate, PBS: polarizer beam splitter, M: mirror, L: lens, SB and RB : signal and reference beam, PRC: photorefractive crystal, HV: high voltage power supply, SLM: spatial light modulator

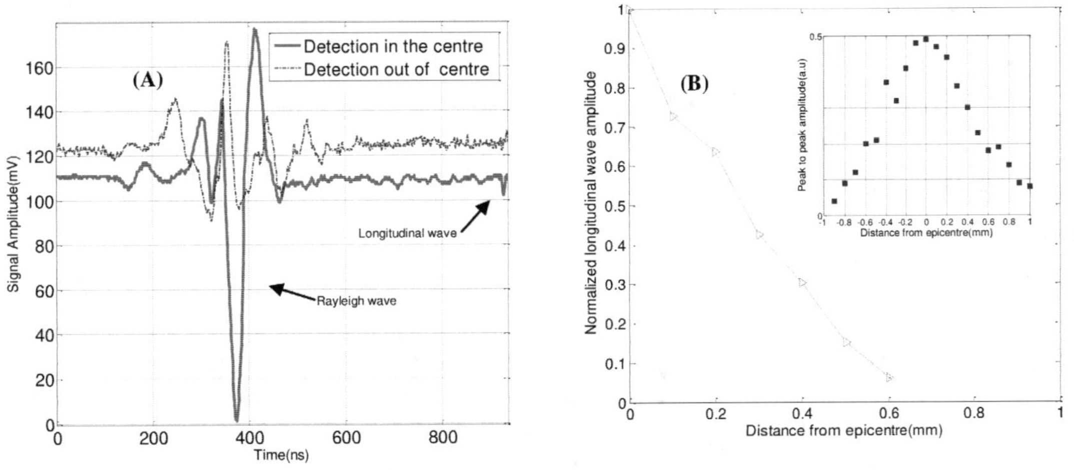

Figure 2: **(A)** Received longitudinal and surface ultrasonic waves by means of adaptive PR interferometer on the surface of a 3 mm thick aluminum sheet: at the center (solid line), 0.5 mm out of the center (dashed line) for a 3 mm diameter full ring generation laser. **(B)**Peak to peak amplitude of the detected bulk waves and surface waves (inset plot) in a ring generation center with 3 mm diameter as a function of distance from the center

In order to study focused bulk waves in the center of the rings, the SLM was programmed to generate 100 rings with diameters between 1 mm to 10 mm on the surface of the sample which is corresponding to focusing angle (the overlapping angle of longitudinal waves on the ring axis) range from 9° to 59°. The focused probe beam of the PR interferometer by means of focusing lenses was adjusted at the annular generation center accurately. To find the ring center, the focused probe beam on the specimen surface using a micrometer stage was scanned horizontally and vertically. As is seen in Fig.2, the Rayleigh wave with higher amplitude in the points very close to the center of the rings achieves. In Fig.3, the detected ultrasonic pulses as a function of ring diameter (1 mm to 9 mm) and time are

Journal of Physics: Conference Series **520** (2014) 012001 doi:10.1088/1742-6596/520/1/012001

displayed. One can see clearly the Rayleigh waves (R) with high amplitude and other detected acoustic waves such as longitudinal and transversal waves.

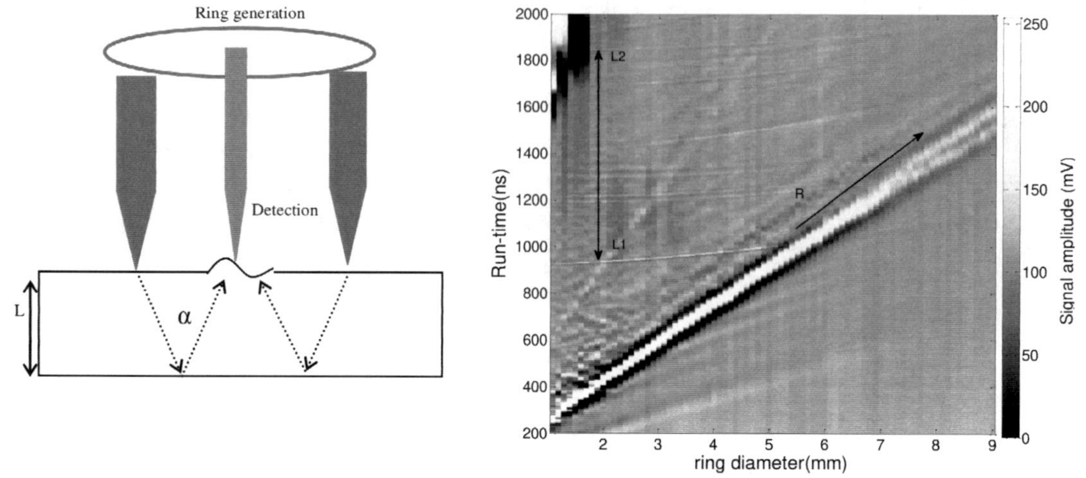

Figure 3: Simple schematic of side view of ring generation and detection in the center on a sample and detected ultrasonic waves at the generated ring center with scanned diameters from 1 mm to 9 mm. L1, L2 and R are longitudinal bulk and Rayleigh waves respectively.

The periodically reflected longitudinal bulk waves (L1, L2) from the backside of the sample with a time of flight of 940 ns are obviously observed and by knowing the ultrasound velocity,6400 m/s, one can estimate the sample thickness. In Fig.4A, the amplitudes of the detected longitudinal bulk waves (L1) in the rings center as a function of focusing angle of the bulk waves (α) inside the sample is displayed. As is seen, as the rings diameter (focusing angle) increases, the amplitude of the longitudinal waves reach to the maximum at the angle of about $\alpha_{max} = 50^\circ$ which is due to convergence of the waves on the central axis points (focal depth) inside the material and near the surface. This angle is close to the directional calculated angle of $\alpha = 60^\circ$ in directivity pattern of longitudinal waves for a thermo-elastic point like laser source in aluminum [1]. J. Guan et al. by using the finite element method in thermo-elastic coupling mode simulated the maximum amplitude of the longitudinal waves for ring-shaped laser illumination on aluminum plates [7]. They calculated that this appears at the angle 53.3° which is in good agreement with our experimental results.

Detection of subsurface defects and the flaws depth with a good resolution is an attractive industrial application of the non-contact ultrasound interferometers. Focusing the bulk waves inside a material in different focal depths by using the ring shape generation method is promising approach to detect and profile internal flaws and defects in materials. To perform such an experiment by use of our BSO PR ultrasonic detector, an artificial hole with 2 mm width in about 1 mm high (D=2 mm from the surface) as a subsurface flaw on the backside of the aluminum plate was drilled. The optimum converging ring was calculated ($D.\tan(\alpha_{max})$) and a ring with about 2 mm radius was generated on the sample surface. The sample was scanned horizontally by using a programmable stage. Fig.4B shows the detected ultrasonic signals of the longitudinal bulk waves and reflected bulk waves from the subsurface defect on the backside of the sample which displays obviously the defect position and its profile.

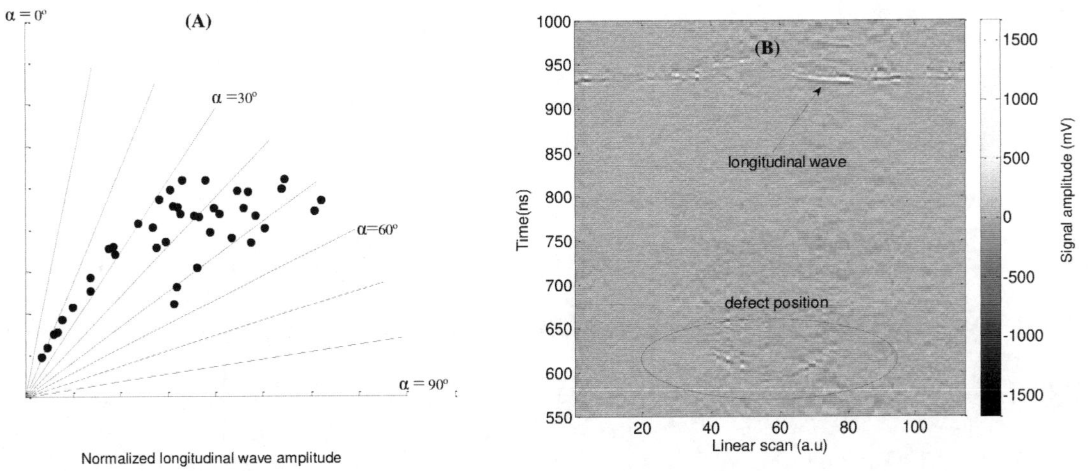

Figure 4: (A) Normalized amplitude to the power density of detected bulk wave in the centre of the rings as a function of focused angle of the bulk waves in a3 mm thick aluminum plate (B) Received ultrasonic longitudinal and reflected bulk signals from the subsurface defect in a 3 mm thick aluminum.

3. Conclusions

We reported that the converging laser ultrasonic wave technique by using the ring-shape laser generation as a contactless and non destructive method has some attractive approaches to evaluate the material with low generation energy in thermo elastic regime and high resolution. Using this technique, the laser generating rings with diameters between 1 to 10 mm and with a 50 μm thickness for full scanning of the aluminum plate surface were generated. Either the focused Rayleigh or bulk waves at the rings center were detected by use of a non-contact BSO PR adaptive interferometer. This promising thermo-elastic method with high resolution can be utilized to evaluate flaws and cracks in composite materials in airframes structure.

References

[1] C. B. Scruby, L. E. Drain, *Laser-Ultrasonics: Techniques and Applications* (Adam Hilger, Bristol, UK, 1990).

[2] C. K. Jen, P. Cielo, J. Bussiere, F. Nadeau and G. W. Farnell, Appl. Phys. Lett. 46, 241(1985).

[3] P. Cielo, F. Nadeau and M. Lamontagne, Ultrasonics, 23, 55(1985).

[4] X. Wang, M. G. Littman, J. B. McManus, M. Tadi, Y. S. Kim, A. Askar, H. Rabitz, J. Appl. Phys. 80, 4274(1996).

[5] M. Clark, S. Sharples, M. Somekh, Meas. Sci. Technol. 11, 1792(2000).

[6] T. Honda, T. Yamashita, H. Matsumoto, Jpn. J. Appl. Phys. 34, 3737(1995).

[7] J. Guan, Z. Shen, X. Ni, J. Lu, J. Wang, B. Xu, Optics and Laser Technology, 39, 1281(2007).

Journal of Physics: Conference Series **520** (2014) 012002 doi:10.1088/1742-6596/520/1/012002

Numerical and Experimental Studies towards Improvements in Laser-Acoustic Microscopy by Optical-Based Sound Beam Shaping

T Windisch
Fraunhofer Institute for Nondestructive Testing, Dresden branch (IZFP-D), Maria-Reiche-Strasse 2, 01109 Dresden, Germany

E-mail: thomas.windisch@izfp-d.fraunhofer.de

Abstract. Incorporating Laser-Ultrasonic systems for excitation and acquisition of ultrasonic waves into NDT&E systems establishes the benefits of contact free methods to high-frequency ultrasonic measurements. Techniques using this huge potential were investigated for a long time but in most cases with focus to surface waves. To study the material bulk properties strong surface damaging laser pulses were mostly used. However, by shortening the pulse length down to some nano-seconds at moderate intensities one can excite ultrasonic waves in the bulk of the material without any damage.
The work presented deals with the prediction of the acoustic fields shape based on arbitrary energy distribution across the excitation laser spot and its use in ultrasonic inspection. An algorithm is discussed which uses the laser beam profile, measured surface normal displacements and bulk wave speeds as input. Comparing simulated an measured data reveals basic agreements but also still some discrepancies caused by additional thermal diffusion in the excitation region. Using this algorithm already offers to analyse the presumed location and shape of longitudinal and transversal wave foci. Further, its capability for optimizing laser-acoustic systems is demonstrated.

1. Introduction

Laser-acoustic techniques offering both, excitation and detection of ultrasonic waves, are still progressing according to the massive improvements of laser sources and systems. Especially in nondestructive testing and materials characterization (NDT&E) laser-acoustic methods receive high attention because of their wide range of application. One major benefit is the ability to handle high frequency ultrasonic waves without the need of any contact or couplant.
Further advantages are found in the wide spectral nature of laser-acoustic sources which build a short pulse-like signal in the time domain. This increases the ability of detecting flaws within the bulk material that are close to other boundaries. Concerning material characterization applications the wide spectrum distribution offers to evaluate frequency dependent parameters within one single measurement.
Beginning in the late 1950s, researchers started to study laser-acoustic sources theoretically and experimentally, wherein ablative and thermoelastic regimes were described [1]. Due to its much stronger acoustic waves high power lasers ablation sources were studied first, although they come with a small dislocation of surfacial matter. In contrast, the thermoelastic regime solely

introduces a temporal-spatial temperature gradient which is preferred by NDT&E. Because of its much weaker acoustic amplitudes, it was hardly possible to detect them by contact methods even less optically, which titled them as non-applicable. Nowadays, compact laser systems offer several thousand pulses per second and to measure displacements in the pico-meter range which dramatically pushed the interest towards thermoelastic sources.
In this article, a method is proposed to calculate thermoelastically generated sound fields in isotropic media excited by different laser beam shapes. It combines an analytic formulation of the source mechanism with spatial and temporal measurements across excitation area. Especially for laser-acoustic microscopy or material characterization it is essential to know, how the waves are radiated, where they superimpose and from which region reflection can be expected. As a result of a better knowledge concerning the sound field, laser-acoustic measurements are dramatically increased in their informative value.

2. Sound fields of arbitrary shaped sources

In order to tailor the sound field to its specific application one need to know the spatial and temporal development of the acoustic waves inside the material. Due to the fact, that it is hardly possible to probe the particles elastic motion in solids, one approach is to sample all necessary information outside of the solid and calculate the sound field as a representation of the real waves. To do so, several mathematical techniques are well known but most of them are limited by assumptions which reduce their applicability to real sources. For example, requirements like source diameters significantly smaller than the acoustic wavelength or the ability to describe the shape in an analytical manner is rarely the case. As shown in figure 1 the intensity distribution across the focal area distinctly differs from mostly point-like assumed shapes. Shown are three intensity distributions across the focal region. The figure on the left represents the so called beam profile of a higher quality cw-laser but not able to excite acoustic waves. In the middle and the right two examples of laser for acoustic excitation having arbitrary shapes. As a consequence, these sources can't be calculated by symmetry-based methods without having significant errors.

2.1. Theoretical Background of thermoelastic sources

While applying a single laser pulse to a free metal surface, part of the radiation is reflected while the remainder is absorbed by the lattice which causes heating. By keeping a power density below $10^6\,\mathrm{W/cm^2}$ at a pulse length in the lower nanosecond range, melting or vaporization of matter is prevented. In time scales of the laser pulse the resulting thermal expansion is limited to a

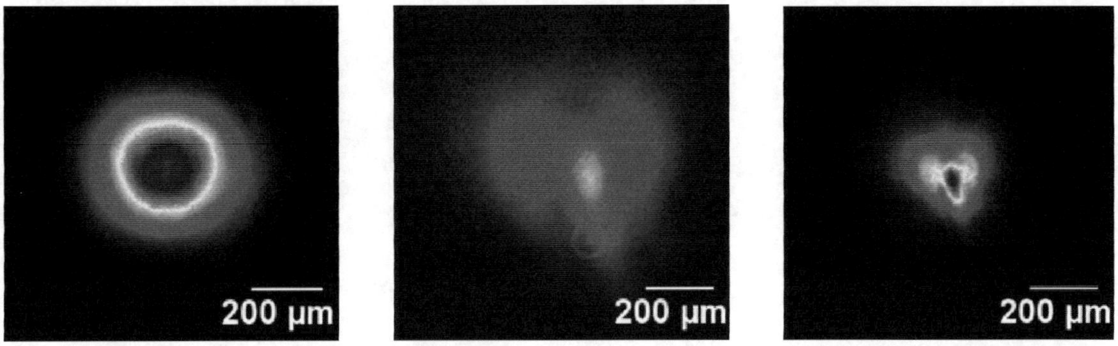

Figure 1. Beam profile measurements representing an almost gaussian laser beam (left) and two intensity distributions of sources for laser-acoustic excitation at large (center) and small focus sizes (right). Bright represents high intensity.

disk-like volume at the surface resulting in a horizontal force dipole [6]. Its thickness equals the thermal skin depth of some μm [2] while the diameter equals the laser spot size. According to the slow thermal diffusion compared to the short pulse duration the temporal characteristic is well approximated by the Heaviside function $H(t)$ as the integral over time of the Dirac pulse $\delta(t)$ [3]. The lattice displacement is directly connected to the surface guided Rayleigh wave as well as the longitudinal and transversal waves propagating into the volume. A diversified discussion concerning the mechanism behind thermoelastic sources can be found in Scruby's standard work *Laser Ultrasonics* [2]. Equations (1) and (2) are taken from this textbook to calculate the angular distribution of longitudinal u_r and transversal wave displacements u_θ. It is important to mention that these directivity patterns are only valid within the Fraunhofer region of a source. Further, they depend only on the materials elastic properties which are represented by the longitudinal and transversal wave speeds c_L and c_T.

$$u_r \propto \frac{\sin 2\theta(k^2 - \sin^2\theta)^{1/2}}{(k^2 - 2\sin^2\theta)^2 + 4\sin^2\theta\,|\cos\theta|\,(k^2 - \sin^2\theta)^{1/2}} \tag{1}$$

$$u_\theta \propto \frac{k \sin 4\theta}{k(1 - 2\sin^2\theta)^2 + 4\sin^2\theta\,|\cos\theta|\,(1 - k^2\sin^2\theta)^{1/2}} \tag{2}$$

Herein, the parameter k is the ratio of longitudinal to transversal wave speed and can also be expressed simply by Poisson ratio ν of the material

$$k^2 = \frac{2(1-\nu)}{1-2\nu} \tag{3}$$

The analytic solutions according to Equations (1) and (2) result in directivity patterns as shown in figure 2 for aluminium ($c_{L,Al} = 6400\,\mathrm{m/s}$, $c_{T,Al} = 3150\,\mathrm{m/s}$) and for steel ($c_{L,St} = 5100\,\mathrm{m/s}$, $c_{T,St} = 3250\,\mathrm{m/s}$). Typical for thermoelastic waves in the far field is the missing radiation at the direction perpendicular to the surface $\theta = 0°$ which is caused by the opposing stresses within the dipole. The transversal wave has a second root at the critical angle $\theta_c = sin^{-1}(1/k)$. Unfortunately analytic formulations are only valid for point-like or symmetric source geometries. In most cases, real sources exhibit arbitrary shapes which can't be handled by these approaches (figure 1). Alternatively, one can utilize FEM software to study the source properties which may easily cause high computational efforts as shown in [7]. To reduce the complexity of simulation work, analytical and experimental data can be combined to perform a hybrid point source synthesis HPSS. By this, information from optics and mechanics are merged. The importance of such linking is also supported by figure 3. In contrast to figure 1 these images show the materials

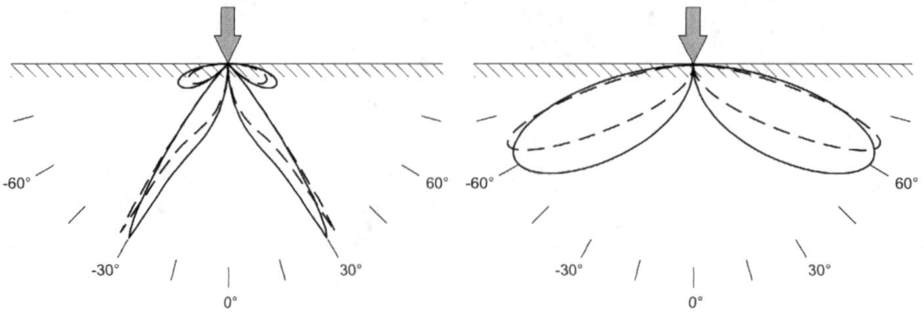

Figure 2. Computed thermoelastic source's directivity of transversal (left) and compression waves (right) at a free surface of aluminum (solid line) and steel (dashed line)

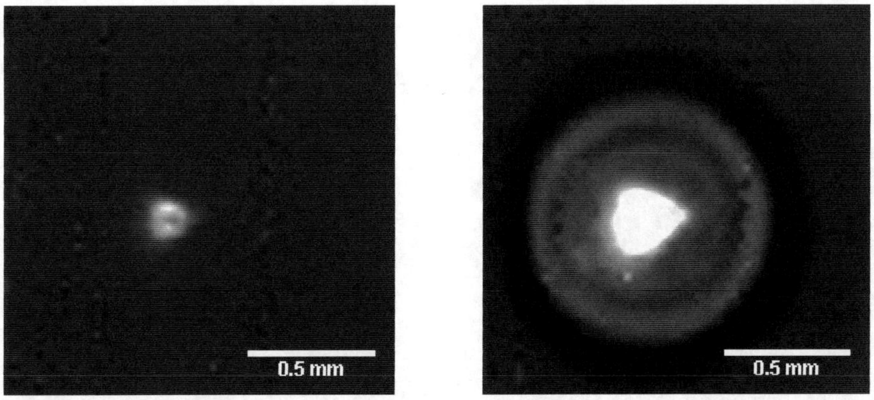

Figure 3. Images of normal surface displacement during thermoelastic excitation as seen by the laser doppler interferometer just after the laser pulse (left) and 200 ns later (right)

mechanical reaction after an optically introduced disturbance. Based on the asymmetric beam profile of the excitation laser a non-uniformly distribution of surface normal displacements are observed which justifies the importance of including the lasers beam profiles in the acoustic design and interpretation of ultrasonic signals.

2.2. Algorithm

As published in [5], the HPSS algorithm combines three main types of input. The first one is the spatial intensity distribution across the source area, acquired from the laser beam profile at the excitation spot. To avoid spatial aliasing, a CCD camera system was utilized having a resolution at least ten times smaller than the smallest acoustic wavelengths considered. The second input is the time resolved surface displacement, as measured by a laser doppler interferometer. By this, no assumptions concerning the temporal properties of the ultrasonic signal are necessary. Finally, a set of material parameters consisting of the wave speeds c_L, c_T and if necessary the wave's attenuation are included. These values are taken from textbooks or measured by subsequent experiments. The algorithm starts to analyse the beam profile and locates a calculation area perpendicular to the surface aligned at the center of the source. Subsequently, each pixel within the beam profile is considered as a single point source. By this, the asymmetric source is divided into point-symmetric sub-sources having the angular intensity distributions according to equations (1) and (2). Furthermore, the transient characteristic is assumed to coincide with the surface displacement, scaled by the pixel's intensity. This is supported by the Poisson ratio, which is linearly linking orthogonal forces, strains or stresses. Finally, a linear superposition of all sub-sources (known as PSS) is used to map the time-resolved 3D wave propagation into the desired 2D area of analysis.

3. Results

All results presented below were taken from experiments performed at the Laser Ultrasonics Laboratory at the Fraunhofer IZFP, Dresden branch. Besides of the examples shown the given conclusions are supported by additional studies concerning variations in frequency and laser spot shape. First we found coincidence in using equation 4 for estimating the bandwidth of the acoustic waves as found in several publications (e.g. [2]). Using the pulse length t_p of the excitation laser, frequencies up to 100 MHz at significant amplitudes are observed.

$$f(-6\text{dB}) \approx \frac{0.1874}{t_p} \tag{4}$$

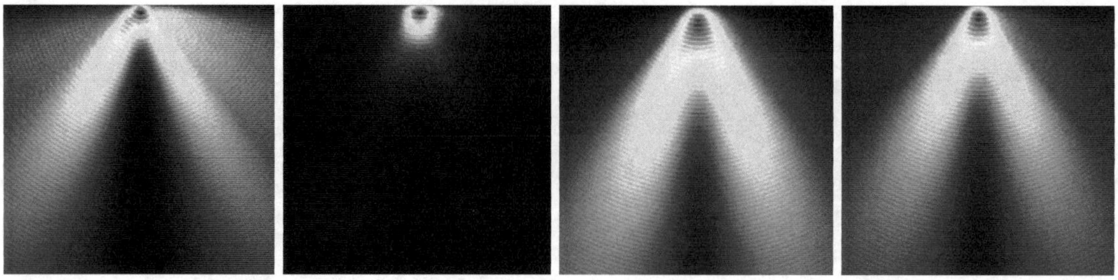

Figure 4. Calculation and spatial decomposition of pressure wave fields excited by a thermoelastic asymmetric source. x- (left) and y-displacement (second left) as orientations parallel to surface, z-displacement (second right) perpendicular to surface and absolute value of all three orientation (right). Each picture is scaled to its maximum value starting at blue for weak up to red for high displacements.

Due to this still high bandwidth as compared to conventional contact based ultrasonic transducer used in NDT, short transient signals can be excited which helps to distinguish closely located scatterers. Analysing the spatial decomposed sound fields reveals additional information about the source properties, which are mostly estimated in a general manner but not in detail as given by such simulation work. In figure 4, four different images are shown which all result from the same asymmetrical thermoelastic source. In these images, the value of each pixel corresponds to the maximum amplitude of the time signal after a Hilbert function based envelope. Besides the absolute value of displacement (most right image) a decomposition into the three spatial orientations was done. All images are scaled to its maximum value. For comparing among themselves the maximum values are $\mathrm{x}_{max} = 0.46$, $\mathrm{y}_{max} = 0.55$ and $\mathrm{z}_{max} = 0.76$ related to $\mathrm{avg}_{max} = 1$. Herein the x direction is orientated parallel to the specimen's surface, the y direction perpendicular to x and again parallel to the surface and finally z perpendicular to the surface. In the most left image, depicting the x component, differences in the distribution of displacement is clearly visible. Two superelevations, one in the left main lobe and a second more widespread in the right side lobe is noticeable. Although the excitation area was kind of a superposition of two tilted elliptic shapes, the distribution of y component (second left) differs dramatically having a single hot-spot close to the surface just below the source. The image corresponding to the z component (second right) is surprisingly almost symmetric. However, the focal region is significantly deeper located than the horizontal orientated one. Finally, the average of all three components is mainly dominated by the strongest z component but has a hot-spot region in a mid-depth and a slightly left shifted distribution of displacements.

A determination of the simulated sound fields accuracy was done by measurements according to a method published in [8]. Using an aluminum alloy sample of 5 mm thickness having wavespeeds of $c_L = 5100\,\mathrm{m/s}$ and $c_T = 2840\,\mathrm{m/s}$ and a stepwise shifting of the excitation location by keeping the detection location the dashed curve was obtained as shown in figure 5. The solid line corresponding to the simulated values was taken from a simulation as discussed earlier but done for a single line at a distance of 5 mm below the surface by using the wave speeds as measured. Comparing simulated and measured values basic analogies like two sidewise shifted maxima or the presence of significant amplitudes at zero degree are observed as well as discrepancies in the angular position of maxima and its symmetry.

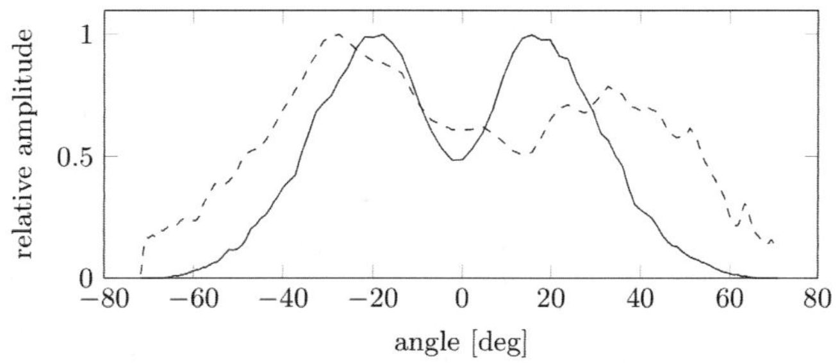

Figure 5. Simulated (solid line) and measured (dashed line) amplitudes over different angles of direction according to a source as shown in fig.1 right image.

4. Discussion and conclusion

Thermoelastic laser-acoustic methods holds both, challenges in simulation or optimization and perspectives in non-contact measurements for NDT&E. As shown here as well as by other researchers simplifications are still necessary to calculate the real sound field in solids. Using hybrid point source synthesis (HPSS) enables to incorporate real spatial source intensity distribution and measured transient displacement signal as a precise replacement for simple assumptions. Common to all methods are their specific restrictions as estimation towards the real physics, which always should be kept in mind during interpreting the results. HPSS incorporates reliable measurements and offers to analyse the different displacement component distributions as an important information for serious interpretation of ultrasonic signals. At the moment, discrepancies between simulated and measured results are still present. Approaches for improvements are seen in incorporating thermal linking between single sources and by replacing the analytic source description by CEFIT simulation of the thermoelastic source [4]. Although at this early stage, the HPSS method was sucsessfully used to improve an laser-ultrasonic setup. Before starting the measurement itself, several intensity distribution of the excitation laser were created and checked versus their sound field in principle. As a result, favourable beam shapes were found and tested at an transmission setup. The corresponding B-scan images of an aluminum sample (thickness 5 mm) having seven drilled holes as artificial defects of 0.7 mm diameter at rising depths are shown in figure 6. Both images were converted to grayscale images, improved in brightness and contrast for better illustration at the same time and cut into single images afterwards. The improvement in SNR and ability to indicate the location of defects is clearly visible. Especially for advanced NDT&E systems, designed for locating scatterers of specific orientation or responding differently to pressure and shear forces, it is inevitable to have

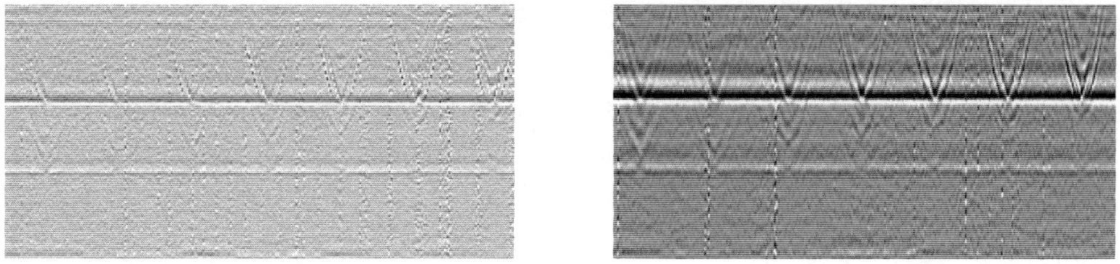

Figure 6. B-Scan transmission images by using raw laser beam (left) and shaped beam after HPSS analysis (right). Horizontal axis location, 70 mm at 250 μm pitch, vertical axis time, 0 to 3 μs starting at bottom line

Journal of Physics: Conference Series **520** (2014) 012002 doi:10.1088/1742-6596/520/1/012002

knowledge about the sensors characteristics. Further by knowing the focal region, unwanted side lobes or the distribution of different wave types one gets able to correctly interpret ultrasonic signals. In a first successful application, HPSS was used to tailor the laser-acoustic source shape of an experimental setup wherein an automated laser-acoustic scan was done for locating defects inside a metallic sample. For both, reflection and transmission B-scan images, the signal to noise ratio could be improved just by changing the energy distribution across the excitation area.

References

[1] G. F. Miller and H. Pursey, *The field and radiation impedance of mechanical radiators on the free surface of a semi-infinite isotropic solid*, Proc. R. Soc. A, Vol. 223, No. 1155, pp. 521-541, The Royal Society, 1954.

[2] C. B. Scruby and L. E. Drain, *Laser ultrasonics techniques and applications*, Taylor & Francis, 1990.

[3] J.F. Ready *Effects of high-power laser radiation*, Academic Press, New York, NY, 1971

[4] F. Schubert, B. Koehler and A. Peiffer, *Time domain modelling of axisymmetric wave propagation in isotropic elastic media with CEFITCylindrical elastodynamic finite integration technique*, Journal of Computational Acoustics, Vol. 9, No. 03, pp. 1127–1146, World Scientific, 2001.

[5] T. Windisch and F. Schubert, *Ultrasonic wave field determination of laser-acoustic sources with arbitrary shapes*,Ultrasonics Symposium (IUS), 2012 IEEE International, pp. = 2184–2186, 2012,IEEE

[6] T. Sanderson, C. Ume and J. Jarzynski, *Longitudinal wave generation in laser ultrasonics*, Ultrasonics, Vol. 35, No. 8, pp. 553 - 561, 1998.

[7] D. Cerniglia, A. Pantano and C. Mineo, *Influence of laser beam profile on the generation of ultrasonic waves*, Applied Physics A: Materials Science & Processing, Vol. 105, No. 4, pp. 959-967, Springer, 2011.

[8] J. M. S. Sakamoto, B. R. Tittmann, A. Baba and G. M. Pacheco, *Directivity measurements in aluminum using a laser ultrasonics system*, Journal of Physics: Conference Series, Vol. 278, 2011, IOP Publishing

Journal of Physics: Conference Series **520** (2014) 012003 doi:10.1088/1742-6596/520/1/012003

A Feasibility Study on Generation of Acoustic Waves Utilizing Evanescent Light

I Matsuya[1], K Matozaki[2], A Kosugi[2] and I Ihara[1]

[1]Department of Mechanical Engineering, Nagaoka University of Technology, Kamitomioka 1603-1, Nagaoka, Niigata, 940-2135 Japan
[2]Graduate School of Engineering, Nagaoka University of Technology, Kamitomioka 1603-1, Nagaoka, Niigata, 940-2135 Japan

E-mail: matsuya@mech.nagaokaut.ac.jp

Abstract. A new approach of generating acoustic waves utilizing evanescent light is presented. The evanescent light is a non-propagating electromagnetic wave that exhibits exponential decay with distance from the surface at which the total internal reflection of light is formed. In this research, the evanescent light during total internal reflection at prism surface is utilized for generating acoustic waves in aluminium and the feasibility for ultrasonic measurements is discussed. Pulsed Nd:YAG laser with 0.36 J/cm^2 power density is used and the incident angle during the total internal reflection is arranged to be 69.0° for generating the evanescent light. It has been demonstrated that the amplitude of the acoustic waves by means of evanescent light is about 1/14 as large as the one generated by the conventional pulsed laser. This reveals the possibility of using a laser ultrasonic technique with near-field optics.

1. Introduction
Laser ultrasonic measurement is a sophisticated technique which enables the evaluation of the mechanical properties of material, crack detection, identification of temperature distribution and so on [1][2]. In this technique, the measurement resolution is limited by the diameter of the incident light spot. Therefore, the novel method that overcomes the resolution limitation in laser ultrasonic measurement is strongly aspired. To extend this technique smaller than the diffraction limitation of the light wave, utilizing the near-field optics is one of the promising alternatives to the conventional laser ultrasonic system. Ultrasonic near-field optical microscopy (UNOM) has been developed to provide nanoscale spatial resolution for ultrasound detection [3]-[6]. The UNOM technique allows local mapping of ultrasound with deep sub-optical wavelength spatial resolution. However, little attention has been given to the point that the generation of ultrasound utilizing near-field optics at local area which is smaller than the wavelength of the light wave. In this article, the evanescent light during total internal reflection at the prism surface is utilized for generating acoustic waves in aluminium plate and the feasibility for ultrasound measurements is discussed.

2. Evanescent wave
The evanescent light is a non-propagating electromagnetic wave that exhibits exponential decay with distance from the surface at which the total internal reflection of light is formed. Figure 1 shows the evanescent light which penetrates inside the material 2. When the total internal reflection of the light is formed, the electric field of evanescent light at the prism surface is defined as follows.

$$E(x,z) = T_0 exp\left(-i\omega t + ik_2 \frac{x}{n}\sin\theta_1\right) \cdot exp\left(-k_2 z\sqrt{\frac{1}{n^2}\sin^2\theta_1 - 1}\right) \quad (1)$$

$$k_2 = \frac{2\pi}{\lambda_2}\ ,\ n = \frac{n_l}{n_h} \quad (2)$$

where T_0 is the maximum amplitude of the electric field, ω is angular frequency of the incident light, k_2 is wave number of the light in material 2, λ_2 is wave length of the light in material 2, and θ_1 is angle of the incident light. In equation (1), if we assume $z = 1/k_2\sqrt{(1/n)^2\sin^2\theta_1 - 1}$, the amplitude of the electric field decreases 1/e as small as the electric field at z = 0. Therefore, the 1/e-penetration depth Λ of evanescent light is defined as shown below [7].

$$\Lambda = \frac{1}{k_2\sqrt{\frac{1}{n^2}\sin^2\theta_1 - 1}} \quad (3)$$

In this research, refractive index of prism n_h is 1.51, refractive index of aluminum n_{al} is 1.37 [8]. According to Snell's law, the critical angle is determined to be 65.1° where the incident light shows the total internal reflection at the interface between prism and aluminum. The critical angle of prism-air interface is determined to be 41.2° as well. In our experiment, the angle of the incident light is arranged at 69.0° which is larger than the critical angle of prism-aluminum interface to generate the ultrasonic waves only by the evanescent light. Figure 2 shows the electric field of evanescent light according to the position from the prism surface to the air [7]. We used Nd:YAG laser of 1064 nm optical wavelength, so the 1/e-penetration depth Λ of evanescent light is determined to be 170 nm by using equation (3). When the evanescent light reaches the aluminum surface, the conduction electrons at the surface screen the interior of the aluminum from the incident light so that the absorption and reflection take place within the surface layer. This 'skin depth' in the aluminum specimen is estimated at 5 nm for the wavelength of ND:YAG laser [9].

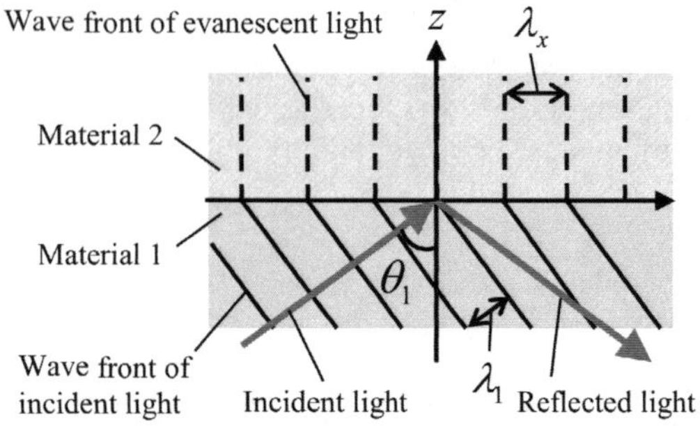

Figure 1. Evanescent light during total internal reflection at prism surface.

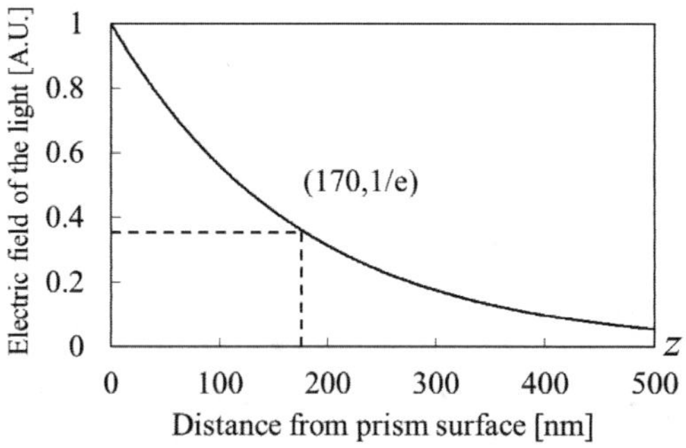

Figure 2. Electric field of evanescent light from prism surface to air.

3. Experimental setup

Figure 3 shows the schematic diagram of experimental setup for measuring the generation of acoustic waves. Figure 4 shows the detailed view of the experimental setup. A well-polished aluminum plate with 5 mm thickness is used as a specimen. The right angle prism is in close contact with the specimen. The evanescent light is formed during the total internal reflection at prism surface by using a pulsed laser. The incident angle of the pulsed laser is determined to be 69.0° which is more than the critical angle of 65.1° derived from Snell's law. Pulsed Nd:YAG laser (maximum power 180 mJ) is operated at 1064 nm optical wavelength, with duration time of 3~5 nsec. A lens with 100 mm focal length is used to focus the pulsed laser onto the specimen. To detect acoustic waves, the ultrasound transducer with coupler is arranged on the other surface of the specimen with proper pressure as shown in figure 4. The ultrasound transducer is longitudinal wave type with 10 MHz band frequency, 9 mm diameter. The power density of pulsed laser is measured by using a power meter to evaluate acoustic waves quantitatively. Laser power is controlled to be 10 mJ, and distance from lens to aluminum is arranged

130 mm which is longer than the focal length of the lens to enlarge the irradiated area on the specimen. Because a damaged region will be introduced inside the prism due to perfectly focused laser, we intended to control the energy density (laser power / irradiated area) of pulsed laser carefully. Although the spot diameter of the laser is usually estimated by $1/e^2$ (13.5%) power as compared with the beam center, we utilized the simple method this time. The irradiated area is estimated by drawing an ellipse over the damaged region on the sensitized paper. The irradiated area is estimated to be 2.75 $\times 10^{-2}$ cm^2, so the energy density is calculated at 0.36 J/cm^2.

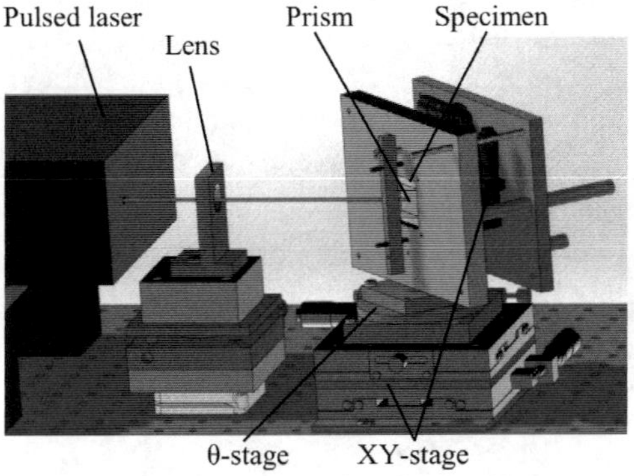

Figure 3. Schematic diagram of experimental setup.

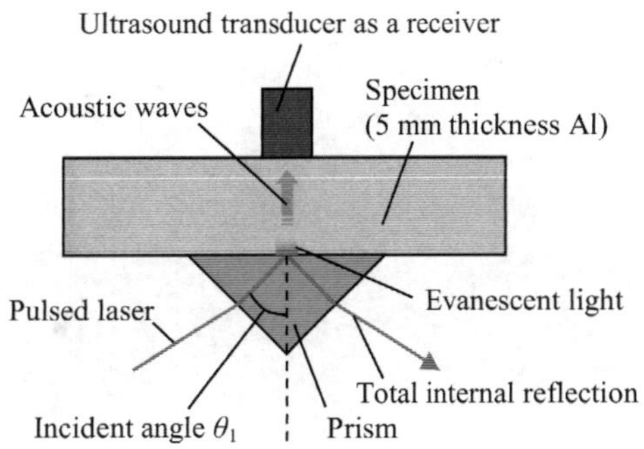

Figure 4. Generation and detection of acoustic waves by evanescent light.

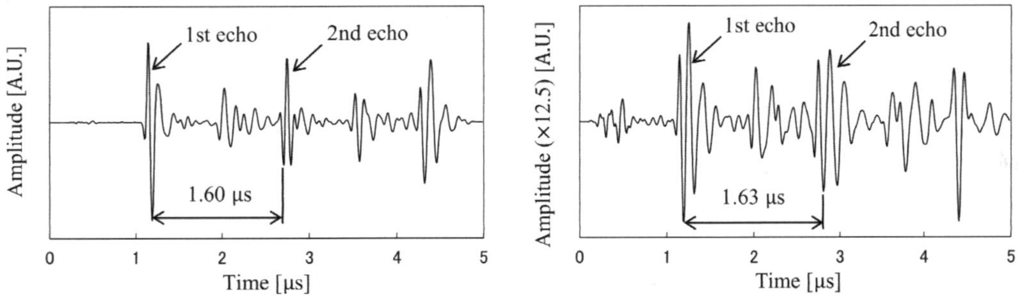

Figure 5. Measured acoustic waves generated by direct laser irradiation (left), evanescent light (right).

4. Results and Discussion

Figure 5 shows obtained acoustic waves generated by conventional pulsed laser and the evanescent light. A 29 dB gain and a 59 dB gain are applied to the original signals generated by the conventional laser ultrasonic method and the evanescent light, respectively. For fair comparison, this gain is canceled after the data processing. The highest signal of the acoustic waves generated by evanescent light is approximately 1/14 as large as the one generated by the conventional pulsed laser. The starting times of both the first and the second echo are almost the same and the time delay from the first echo to the second echo is estimated at 1.63 μs utilizing cross correlation method. Therefore the velocity of ultrasound was estimated at 6130 m/s. This value agrees well with the reference value of longitudinal wave velocity in aluminum (6260 m/s). These results indicate that it is possible to generate acoustic waves by means of evanescent light.

The condition of contact between the prism and the specimen which affects irradiation degree of evanescent light is examined. The surface roughness of prism and aluminum specimen is investigated by AFM. Figure 6 shows the representative surface of a prism and specimen where both materials contact each other and evanescent light is formed. The maximum distance from the highest to the lowest point is 44 nm at the prism surface and 153 nm at the aluminum specimen, respectively. The surface accuracy (surface height differential on whole surface) of the prism is 158 nm according to the vender specifications. The surface accuracy of the aluminum specimen for the contact area is 1362 nm investigated by laser microscope scanning. The sum of two surface roughnesses (197 nm) is almost the same as the 1/e-penetration depth of evanescent light (Λ = 170 nm) in equation (3). However, the irradiation degree of the evanescent light depends on the surface accuracy of the aluminum specimen because of the existence of the spatial gap between the prism and the aluminum specimen. It is considered that the relatively large gap causes the very weak signal generated by evanescent light in this experiment. Due to the existence of the average gap between the prism and the specimen, the actual penetration depth of the evanescent light in the air is estimated to be reduced by an average of 450 nm when we consider the signal degradation (1/14) and the relationship between the electric field of evanescent light and the distance from the prism as shown in Fig. 2. Note that there are some contact points at the actual prism-specimen interface as with some gaps.

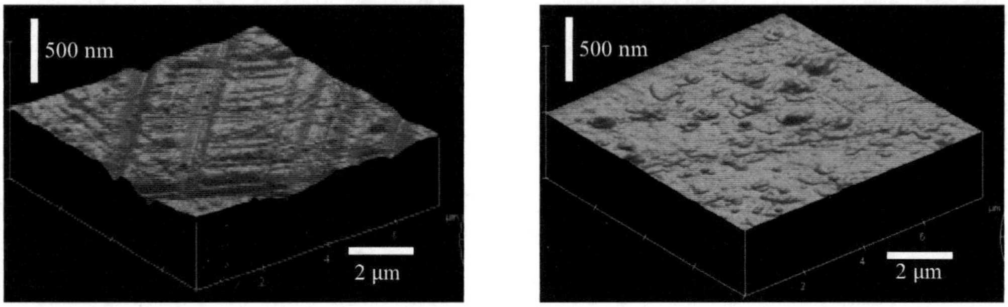

Figure 6. Surface roughness of aluminum specimen (left), prism (right) measured by AFM.

5. Conclusion

A novel method generating acoustic waves by evanescent light during total internal reflection at prism surface is presented. Nd:YAG laser with 0.36 J/cm^2 power density is used and the incident angle is determined to be 69.0° which is more than the critical angle of the total internal reflection for generating the evanescent light. It has been demonstrated that the intensity of the acoustic waves by means of evanescent light are about 1/14 as large as the one generated by the conventional pulsed laser. The time delay from the first echo to the second echo is approximately 1.63 μs, which almost agrees with the expected value determined from the longitudinal velocity of the aluminium by conventional pitch-catch method. These results reveal the possibility of using a laser ultrasonic technique with near-field optics.

Acknowledgments

This work was supported by grant-in-aid for young scientist B (25820005) and grant-in-aid for scientific research B (25289238) of KAKENHI.

References

[1] Kosugi A, Ihara I and Matsuya I 2012 *Jpn. J. Appl. Phys.* **51** 07GB01
[2] Hayashi T, Murase M and Salim M N 2009 *J. Acoust. Soc. Am.* **126** 1101
[3] Vertikov A, Kuball M, Nurmikko V and Maris H J 1996 *Appl. Phys. Lett.* **69** 2465
[4] Blodgett D W and Spicer J B 1999 *Appl. Phys. Lett.*, **74** 3735
[5] Hillenbrand R and Keilmann F 2002 *Appl. Phys. Lett.* **80** 25
[6] Ahn P, Ahang Z, Sun C and Balogun O 2013 *J. Appl. Phys.*, **113** 234903
[7] Ohtsu M and Hori M 1999 *Near-Field nano-optics* (Plenum Pub Corp)
[8] Rakic A D 1995 *Applied Optics* **34** 4755
[9] Scruby C B and Drain L E 1990 *Laser Ultrasonics Techniques and Applications* (Taylor & Francis Group)

Journal of Physics: Conference Series **520** (2014) 012004 doi:10.1088/1742-6596/520/1/012004

The SKED: speckle knife edge detector

SD Sharples, RA Light, SO Achamfuo-Yeboah, M Clark and MG Somekh

Electical Systems & Optics Research Division, Faculty of Engineering, University of Nottingham, University Park, Nottingham NG7 2RD, UK

E-mail: `steve.sharples@nottingham.ac.uk`

Abstract.

The knife edge detector—also known as optical beam deflection—is a simple and robust method of detecting ultrasonic waves using a laser. It is particularly suitable for detection of high frequency surface acoustic waves as the response is proportional to variation of the local tilt of the surface. In the case of a specular reflection of the incident laser beam from a smooth surface, any lateral movement of the reflected beam caused by the ultrasonic waves is easily detected by a pair of photodiodes. The major disadvantage of the knife edge detector is that it does not cope well with optically rough surfaces, those that give a speckled reflection. The optical speckles from a rough surface adversely affect the efficiency of the knife edge detector, because 'dark' speckles move synchronously with 'bright' speckles, and their contributions to the ultrasonic signal cancel each other out. We have developed a new self-adapting sensor which can cope with the optical speckles reflected from a rough surface. It is inelegantly called the SKED—speckle knife edge detector—and like its smooth surface namesake it is simple, cheap, compact, and robust. We describe the theory of its operation, and present preliminary experimental results validating the overall concept and the operation of the prototype device.

Optical detection of ultrasound is achieved by measurement of the change in surface displacement, velocity, or angle [1, 2]. The two principal ways to detect these changes are using knife edge detectors (KEDs)—a technique also known as optical beam deflection—and interferometers. The KED is a particularly attractive device on account of its low cost and simplicity. As the sound wave propagates under the incident optical beam the reflected beam tilts backwards and forwards at the frequency of the ultrasonic waves. The optical beam is detected in a split detector, as in figure 1, so that in part of the cycle the signal in part A is greater than part B and in the other part of the cycle the signal B is greater than signal A. The outputs of the two detectors—either in terms of current or voltage—are subtracted one from the other using a differential amplifier, and the result is a signal proportional to the local tilt of the surface under the incident optical beam.

This works very nicely on optically smooth surfaces, where there is a single, specular reflection. On optically 'rough' surfaces, variations in the height of the sample at the laser beam focus cause different parts of the reflected beam to have different phase components, leading to local areas of constructive and destructive interference. This is known as speckle. This causes problems for the knife edge detector as the whole speckle pattern moves synchronously with the ultrasonic wave, assuming the amplitude is considerably lower than the acoustic wavelengths—in practice, true for all ultrasonic inspection. This means that the contributions from the 'bright' speckles moving from one photodiode to the other are cancelled out by the contributions from the 'dark'

Journal of Physics: Conference Series **520** (2014) 012004 doi:10.1088/1742-6596/520/1/012004

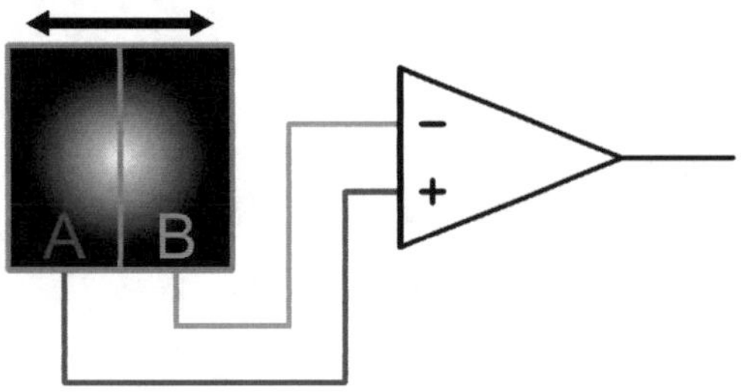

Figure 1. Principle of the conventional knife edge detector (KED). The difference between the electrical currents, proportional to the incident light in each half of the split detector, is determined by the difference amplifier.

speckles moving in the same direction. The result is a rapid decline in the signal to noise ratio of the signal with increasing surface roughness.

Some of the interferometric detection techniques are either inherently tolerant to speckle—such as the Fabry-Pérot interferometer [3]—or can be made to adapt to the speckle—such as those based on two wave mixing [4], or random quadrature demodulation [5]. In this paper, we describe how the optical beam deflection technique can be modified to be tolerant to optical speckle from rough surfaces. It is achieved by a novel design of an active adaptable photodiode pixel array sensor. It is inelegantly called the SKED—speckle knife edge detector—and like its smooth surface namesake it is simple, cheap, compact, and robust. We first describe the theory of its operation, and then present preliminary experimental results validating the overall concept and the operation of the prototype device.

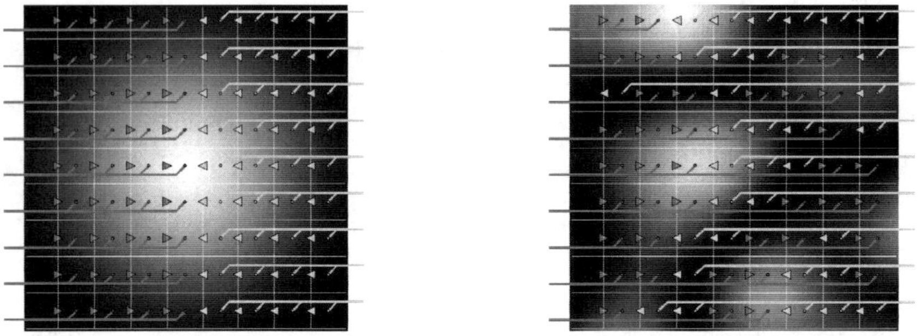

Figure 2. Schematic showing the operation of (left) the conventional knife edge detector, and (right) the speckle knife edge detector. The intensity gradient between adjacent pixels defines the output of one of those pixels.

The principle of operation of the SKED is illustrated in figure 2. The left hand image shows the reflected beam from a smooth surface so as this beam moves from left to right more signal is detected on the 'green' side compared to the 'red' side, exactly the same as in figure 1 above. When the light is reflected from a rough surface light breaks down into a speckle pattern with dark and light areas, moreover, its spatial extent becomes much greater. When the sound wave passes under the rough surface the speckle pattern will oscillate but the signal will not change on average since the light has been scrambled and there is no indication whether the left hand side signal will increase or decrease. If, however, we focus on a single bright speckle we will see it oscillate backwards and forwards giving periodic signal changes. The trick then is to redefine the regions where in the speckle pattern and feed them into the difference amplifier of figure 1. The SKED unravels the speckle by measuring the differences in light intensity across the detector array. Between pairs of adjacent pixels, the chip calculates whether there is more light falling on the left, or the right (represented by the left and right arrows of figure 2). If there is more light falling on the right, then the output from the right hand pixel is sent one way (coloured

red); if there is more light falling on the left, then the output from the right hand pixel is sent the other way (coloured green). If we subtract the sum of all the green pixels from the sum of all the red pixels, we get a signal proportional to the horizontal movement of the speckles, and hence of the local tilt of the sample surface caused by the ultrasonic wave.

The output is now the same as the conventional KED. Measurement of the intensity changes has simply allowed us to use the light reflected from the rough sample and reorganise it so that it behaves as if the reflection were from a smooth surface. The sensor configures its own architecture in response to the conditions of the reflected light. If the sound wave is propagating vertically rather than horizontally a conventional KED is insensitive to this motion, but the SKED allows us to address this by allowing the user to select whether to compare adjacent pixels horizontally or vertically.

The SKED device is a CMOS (complementary metal oxide semiconductor) integrated circuit, fabricated using a standard 0.35 μm 4 metal layer process. The device consists of a 32×32 array of pixels, plus some ancillary routing circuitry to send and receive digital and analogue signals off-chip. Each individual pixel consists of a photodiode which absorbs the incident laser light reflected from the sample under investigation, plus some digital logic and analogue components for mirroring and comparing the photocurrent. The fill factor is 54%. A simplified conceptual schematic is shown in figure 3.

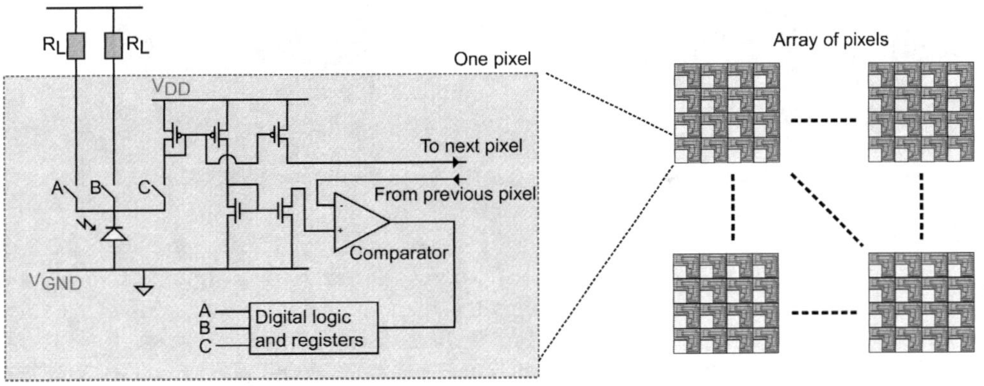

Figure 3. Conceptual schematic showing a single SKED pixel (left) and how this fits into the array of pixels in the SKED integrated circuit.

The SKED has two modes of operation: *configure* and *experiment*. In *configure* mode, all the pixels are isolated from each other—switches 'A' and 'B' in figure 3 are open, switch 'C' is closed—which allows the photocurrent of all adjacent pixels to be compared. The output of the comparator is used to determine whether one of each pair of pixels will be connected to channel A or B—this is actuated when the chip is put into *experiment* mode. At this point, all pixels are connected to one of the two analogue output channels.

In normal operation, the mode of operation would be synchronised to the ultrasound excitation source. Just prior to the point at which ultrasound needs to be detected, the SKED would be briefly put into *configure* mode, whereupon it would adapt to the optical speckle, before being returned to *experiment* mode which would allow the ultrasonic signal to be detected.

The device can also be programmed, in that the output channel of each pixel can be determined by the user. A programmable integrated circuit (PIC) provides an interface between the SKED IC and a USB connection to a host PC. This interface can also be used to read which output channel is active on each pixel.

Figure 4 illustrates the improvement that can be achieved in terms of signal consistency and

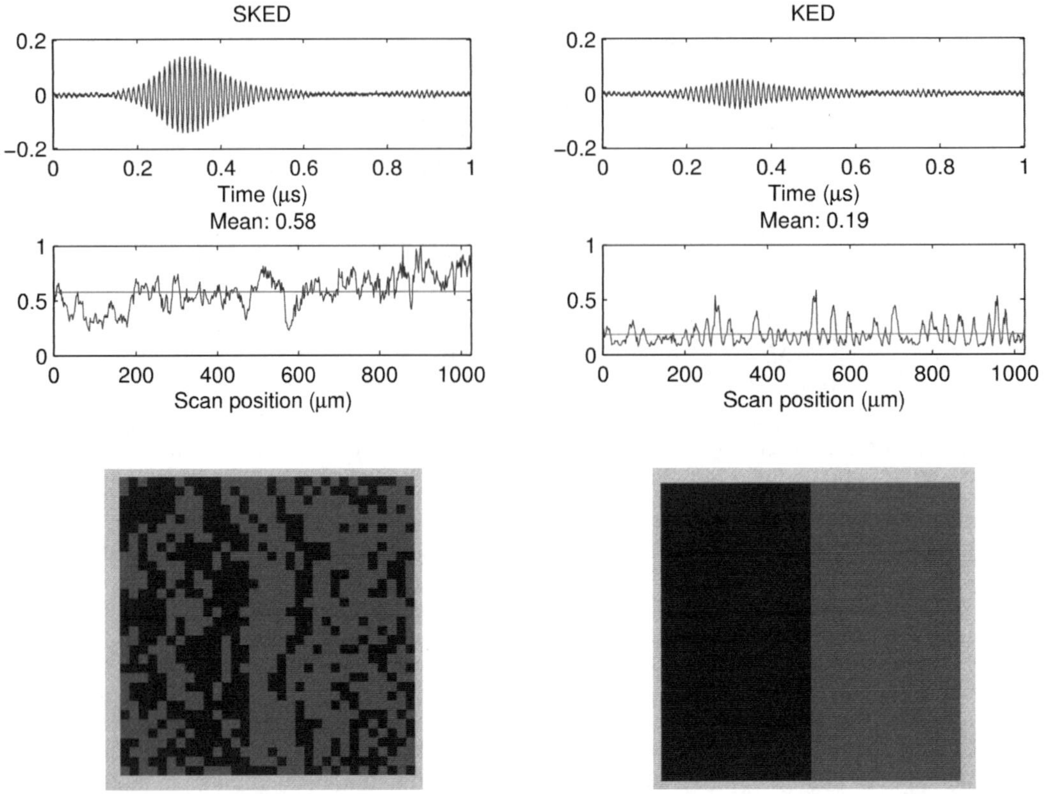

Figure 4. On the left, the SKED device automatically configues itself as a line scan is performed on machined aluminium; on the right, the SKED device is manually configured to behave as a conventional KED. The upper traces are the ultrasonic signals at the last point in the line scan. The centre traces show the mean signal level, normalised to the maximum of all recorded signals. The images at the bottom represent the configuration of the pixel outputs for the last point in the line scan.

amplitude (with respect to fixed noise levels) when the SKED device operates in self-adapting mode, compared to when it is manually configured to operate as a conventional KED with two fixed photodiodes. The ultrasonic signal is a laser-generated surface acoustic wave (SAW) packet with a centre frequency of 82MHz, on an aluminium sample with a machined surface. The theoretical bandwidth of the prototype device is 16MHz; this is limited by the capacitance of the photodiodes and the resistance of the transmission gates used to switch the output of each pixel to channel A or B. Despite this, the device is capable of detecting thermolastically excited SAWs at 82MHz, albeit with a first order filter roll-off of 20dB/decade.

We will fully characterise the device shortly and present the results in a future paper.

References

[1] Monchalin J P 1986 *IEEE Transactions on Ultrasonics, Ferroelectrics and Frequency Control* **33** 485–499
[2] Murfin A S, Soden R A J, Hatrick D and Dewhurst R J 2000 *Measurement Science and Technology* **11** 1208–1219
[3] Monchalin J P 1985 *Applied Physics Letters* **47** 14–16
[4] Ing R K and Monchalin J P 1991 *Applied Physics Letters* **59** 3233–3235
[5] Pouet B, Breugnot S and Clemenceau P 2006 *AIP Conference Proceedings* **820** 233–239

Development of a Fibre-Phased Array Laser-EMAT Ultrasonic System for Defect Inspection

C Pei[1,2], K Demachi[1], T Fukuchi[3], K Koyama[1], M Uesaka[1] and Z Chen[2]

[1]School of Engineering, The University of Tokyo, Tokyo 113-0032, Japan
[2]School of Aerospace, Xi'an Jiaotong University, Xi'an 710049, China
[3]Electric Power Engineering Research Laboratory, Central Research Institute of Electric Power Industry, Yokosuka-Shi, Kanagawa 240-0196, Japan

Email: peicx2009@gmail.com

Abstract. In this work, a phased array laser ultrasound system with using fibre optic delivery and a custom-designed focusing objective lens has been developed for enhancing the ultrasound generation. The fibre-phased array method is applied to improve the sensitivity and detecting ability of the laser-EMAT system for defect inspection.

1. Introduction

The laser-generated ultrasound has been widely studied in various ultrasonic testing techniques, such as laser ultrasonic [1-3] and laser-EMAT ultrasonic [4, 5]. And, it is known that the ultrasonic wave can be generated in thermo-elastic regime or in ablation regime when laser power density is less or larger than the material melting threshold. However the potential ablation damage on the material surface can be minimized in the thermo-elastic regime, one of the principle disadvantages of laser ultrasonic is the low signal strength for thermo-elastically generated ultrasound. To overcome this problem, one of the most efficient ways is using the phased-array (PA) method with a multi-beam laser source [6, 7]. Besides, the PA method can also be used to steer the ultrasonic wave into the area which we are interested on. Until now, two methods of PA laser ultrasound have been mainly reported. One is based on using an optical fiber phased array [6]; and the other one is based on using an array of laser sources [7]. The advantage of using fibers over multiple laser sources is that the fiber is much cheaper, portable and flexible, and only a single laser is needed to provide multiple sources.

A combination of laser ultrasonics with time-of-flight diffraction (TOFD) measurement has been already reported in a few literatures [2, 3]. However, in those works, the ultrasound in ablation regime was used in the measurement. In this work, a PA laser ultrasound system with using fiber optic delivery and a custom-designed focusing objective lens is developed to improve the sensitivity and detecting ability for defect inspection by enhancing the ultrasound generation. The enhanced laser ultrasound in thermoelastic regime is studied for TOFD measurement of cracks in different specimens without ablation damage.

2. Experimental system and method

2.1. Overall system design and setup

As shown in Fig. 1, the phased array laser ultrasound system was constructed with a Q-switched Nd:YAG laser (providing pulses with duration of 10 ns at a repetition rate of 10 Hz) and a three-source fiber-phased array. Three laser beams from the two beam splitters, each with pulse energy up to 10 mJ was coupled into three 400 μm diameter silica-silica fibers with fiber couplers. The length of the three fibers measure are 2.5 m, 42.5 m and 82.5 m, respectively, resulting in a time delay of 0.194 μs. The distal ends of the fibers are connected with a custom-designed focusing objective lens to generate a laser array (three line sources with given dimensions) on the surface of a specimen. The width and length of each laser array element focused on the specimen surface are around 0.8 mm and 10 mm, respectively. The separation between each element is about 1.5 mm. Therefore, a series of ultrasound pulses generated by each laser array element at slightly different time would combine together to add energy at a steering direction. The enhanced ultrasonic waves generated by the laser array were received by an electromagnetic acoustic transducer (EMAT) that was mounted about 0.5 ± 0.1 mm above the specimen surface. Two EMATs, an out-of-plane EMAT and an in-plane EMAT [8], were applied for the ultrasound detection.

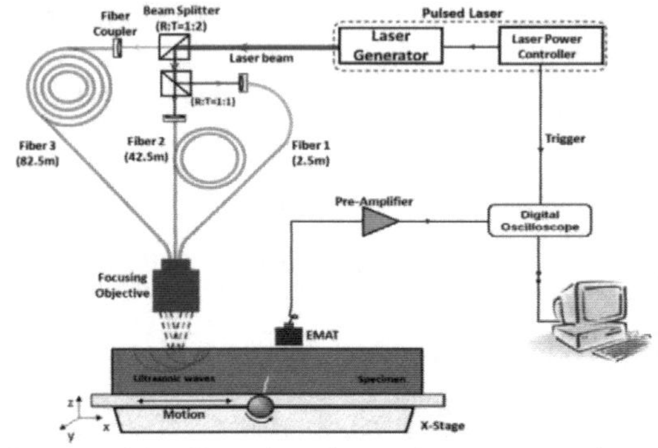

Fig. 1 Schematic diagram of the fiber-phased array laser-EMAT ultrasonic testing system

2.2. Measurement method

The schematic diagram of measurement of cracks with using PA laser ultrasound and EMAT detection by TOFD method is shown in Fig. 2. Unlike the conventional symmetrical TOFD layout, a laser beam and an EMAT (in-plane EMAT or out-of-plane EMAT) in an unsymmetrical arrangement are used as a generator and detector, respectively, in this work. The main reason of this modification is that the out-of-plane EMAT detector has the best sensitivity to the longitudinal wave (and the in-plane EMAT detector has the best sensitivity to the shear wave) propagating perpendicularly to the receiving area. Besides, in the modified setup, the diffraction wave has shorter travelling path. So it can be detected in larger amplitude. Therefore, a much stronger diffraction signal can be acquired when the EMAT is just above the crack. The ultrasonic transmission modes and paths are also shown in the figure. With laser-EMAT ultrasonic, it is possible to accurately measure the absolute propagation time of the different waves. Therefore, the depth of the crack tip in the specimen can be obtained by either Eq. (1) or Eq. (2):

$$d = C_L \cdot \frac{t_{LTL}^2 - t_{La}^2}{2t_{LTL}} \tag{1}$$

$$d = C_L C_S \cdot \frac{C_L t_{LTS} - \sqrt{C_S^2 t_{LTS}^2 + (C_L - C_S) t_{La}^2}}{C_L^2 - C_S^2} \tag{2}$$

where d is the depth of the crack top tip in the specimen, C_L the velocity of L wave, C_S the velocity of S wave, and t_{La}, t_{LTL} and t_{LTS} are the time of flight of La, LTL and LTS, respectively. LTL is a crack-tip-diffracted longitudinal wave and LTS is a longitudinal-shear mode-converted wave diffracted on the crack tip. Therefore, both LTL and LTS can be applied for the TOFD measurement of cracks. LBL is the reflection wave at the crack base (travelling along the crack before being radiated at the tip) and LBS is the mode-converted wave at crack base.

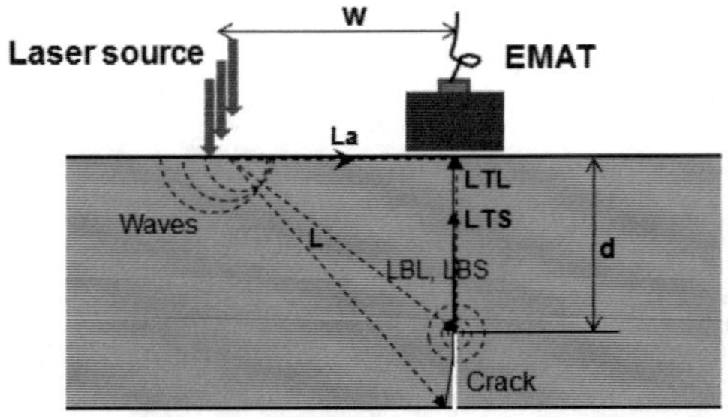

Fig. 2 Schematic diagram of ultrasonic transmission modes and path for crack measurement

3. Measurement results

In order to validate the performance of this system for defect inspection, a measurement of inner-surface cracks in a stainless steel specimen (SUS304) and an aluminum (99.3%) specimen by TOFD method has been studied. Each specimen was a plate (200mm×80mm and 20 mm thick) with three electrical discharge machining (EDM) cracks at the rear side (with positions of 50 mm, 100 mm and 150 mm). The three cracks with a width of 0.2 mm and heights of 10 mm, 5 mm and 2 mm, respectively, were induced entirely through the specimen width. During the experiment, it is found that the measured signal amplitude of LTL diffraction wave is much larger than that of LTS wave in the steel specimen. Conversely, the measured signal amplitude of LTS wave is much larger than that of LTL wave in the aluminum specimen.

Fig. 3 shows the A-scan signals of TOFD measurement of the 10 mm crack in the steel specimen with using a single laser beam and three PA laser beams, respectively. In this measurement, an out-of-plane EMAT was used as the receiver with a separation distance of 28 mm from the laser source. It can be observed that LTL diffraction wave from the crack tip does not present in the waveform generated by a single laser beam, but can be identified in the waveform generated by three PA beams.

Fig. 4 shows the A-scan signals of TOFD measurement of a 10 mm crack in the aluminum specimen with using a single laser beam and three PA laser beams, respectively. In this measurement, an in-plane EMAT was used as the receiver with a separation distance of 22 mm from the laser source. As an in-plane EMAT was used here, the crack diffraction signal LTS can be identified in the waveforms. Besides, as aluminum is a more conductive material, much stronger signals can be received by the EMAT. It can also be observed that the signal-to-noise ratios (SNRs) of LTS signal with three PA laser sources is significantly improved. It is also can be observed in Figs. 3 and 4 that the SNRs in aluminum specimen are much larger than that in steel specimen. This is because EMAT detectors have larger sensitivity in aluminum which is more conductive than steel.

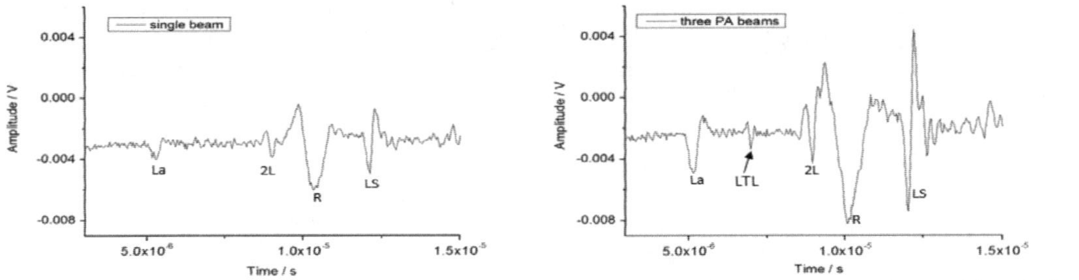

Fig. 3 A-scan signals picked up by an out-of-plane EMAT for 10 mm crack in the steel specimen with a single laser beam, and three PA laser beams

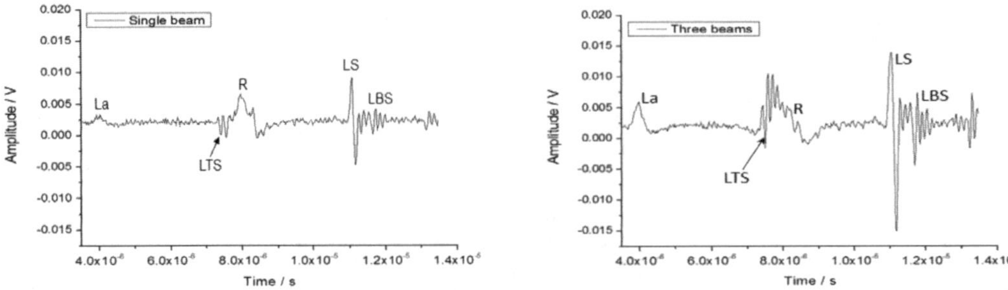

Fig. 4 A-scan signals picked up by in-plane EMAT for 10 mm crack in the aluminium specimen with a single laser beam, and three PA laser beams

Fig. 5 shows the B-scan images of the TOFD measurement by scanning the steel specimen and aluminium specimen, respectively, with using three PA laser beams. The scanning position x is the position of EMAT detector. Beside the lateral wave La, longitudinal echoes 2L and LBL, Rayleigh wave R, mode-converted echo LS, the three diffracted signals LTL_1, LTL_2 and LTL_3 induced by the 10 mm Crack, 5mm Crack and 2 mm Crack, respectively, in the stainless steel were received by the out-of-plane EMAT. The three diffracted signals LTS_1, LTS_2 and LTS_3 induced by the 10 mm Crack, 5mm Crack and 2 mm Crack, respectively, in the aluminium specimen were received by the in-plane EMAT. All these crack signals can be identified clearly.

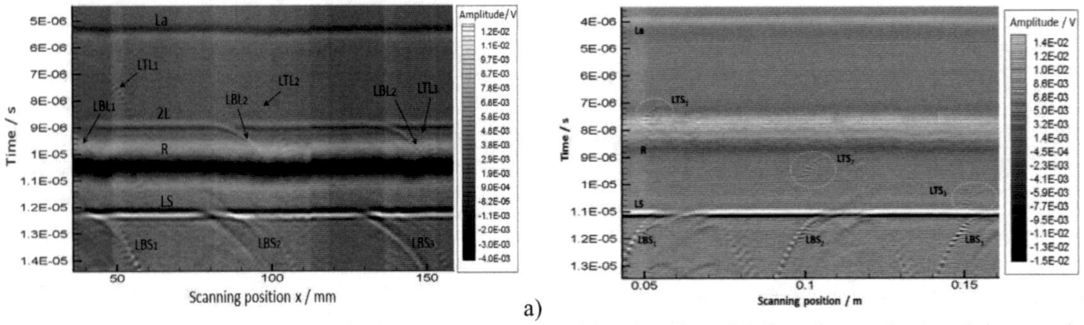

a) b)

Fig. 5 Result of B-scan image of TOFD measurement with using three PA laser beams in a) stainless steel specimen, b) aluminum specimen

As shown in Fig. 4 and Fig. 5, the time of flight t_{La}, t_{LTL} or t_{LTS} for the cracks can be measured, respectively. In Fig. 5, it can be observed that the EMAT is just above the cracks at the center of the diffraction signal patterns. And the diffracted signal has relatively larger strength when the EMAT detector is just above the crack. According to the modified TOFD layout, the depth of crack tip in the specimen can be solved by Eq. (1) or Eq. (2). For the crack existing at the rear side of the specimen, the crack height h can be acquired with: h=H-d, where H is the thickness of the specimen. Table 1 shows the results of crack height evaluation in the stainless steel specimen and aluminum specimen, respectively. It can be observed that all the measurement results are very close to the actual ones.

Table 1 Results of crack height evaluation in steel and aluminium specimens

Actual crack height / mm	Measured height in steel specimen / mm	Measured height in Al specimen / mm
2	2.25	2.26
5	5.40	5.20
10	9.72	9.81

4. Conclusions

A phased-array laser ultrasound system using optical fiber delivery and a custom-designed focusing objective lens has been developed for enhancing the ultrasound generation. The enhancement of crack inspection by using phased array laser ultrasound is validated by experiment. This system was successfully applied to measure the inner-surface cracks both in the stainless steel specimen and aluminum specimen with TOFD method. The diffraction signals from the three EDM cracks with heights from 2 mm to 10 mm can be clearly identified both in the two kinds of specimen. The heights of the three cracks are precisely evaluated from the time of flight of the diffraction waves. It shows that the fiber-phased array laser-EMAT ultrasonic is a very promising non-contact approach for defect measurement.

References

[1] C. B. Scruby and L. E. Drain 1990 Laser Ultrasonics: Techniques and Applications
[2] Q. Shan and R. J. Dewhurst 1993 Applied Physics Letters vol 62 p 2649
[3] T. Mihara, Y. Otsuka, H. Cho, K. Yamanaka 2000 Review of Progress in Quantitative Nondestructive Evaluation vol 19 p 255
[4] D. A. Hutchins, F. Hauser and T. Goetz 1986 IEEE Trans. Ultrason. Ferroelectr. and Freq. Cont. vol 33 p 478
[5] P. R. Murray, R.J. Dewhurst 2002 J. Ultrasonics vol 40 p 771
[6] Bruinsma, A. J.A., and Vogel J.A. 1988 Applied Optics. Vol 27 p 4690
[7] T.W. Murray, J.B. Deaton Jr., J.W. Wagner 1996 J. Ultrasonics vol 34 p 69
[8] B. Dutton, S. Boonsang, and R. J. Dewhurst 2006 Sens. Actuators A vol 125 p 249

Graphitized Layer Buried in a Diamond: SAW Generation under Picosecond Optical Excitation

A Klokov, V Tsvetkov, A Sharkov, D Aminev and R Khmelnitskiy

P.N.Lebedev Physical Institute, Moscow, Russia

klokov@sci.lebedev.ru

Abstract. The excitation of SAW was detected on diamond with built-in ion-implanted graphitized layer under its illumination with femtosecond laser pulses. The spectral width of a SAW pulses were in the range of 1.5÷2 GHz. It was found out that the anisotropy of the SAW propagation was practically absent. The increase in the implantation dose from $4\cdot10^{15}$ cm^{-2} to $12\cdot10^{15}$ cm^{-2} was shown to give rise to a dispersion of SAW propagation.

1. Introduction

Diamond possesses a unique combination of elastic, thermal and electrical properties that makes it an exceptionally attractive material for MEMS as well as for SAW signal processing devices operating in gigahertz range [1]. Formation of various periodical structures is usually required for SAW devices implementation; these periodical structures reflect or simply change the SAW propagation conditions. Such structures can be realized on a basis of ion-implanted graphitized layers build-in the diamond matrix [2]. The information about elastic properties of such layers, e. g., on the sound velocity and attenuation is of high importance. Earlier we have found out that under picosecond optical excitation the bulk coherent phonons are generated by such layers [3]. Depending on the graphitized layer thickness (160-290 nm) and its depth (230-300 nm) the phonons with frequencies in the range of 20 - 120 GHz are generated [4].

This work is devoted to investigations with the help of two-colors picoseconds interferometric method [5] of SAW generation and propagation in structures based on diamond with build-in graphitized layers.

2. Experiment

The buried graphitized layers were created by implanting C^+ ions with the energy of 350 keV into a IIA-type ~300 μm thick natural diamond wafer through a set of masks. Implantation doses are shown in Table 1. The implantation was followed by one-hour vacuum annealing at ~1600°C. The depth of location and thickness of the layers were defined by optical methods similarly to [6] that are shown in Table 1.

The complex refractive index of the layer was n_{Gr}=2.2+j0.88. The reflected light color of those areas with different implantation doses changed from green to dark blue, which was accounted for the interference in optical wavelength range. This allowed us to assume optical perfection of the diamond/graphitized layer interface [7]. The absorption coefficients α of the areas with implanted layers at the wavelength of 400 nm are shown in Table 1.

Table 1. Implantation doses and parameters of buried graphitized layers.

Dose, $\times 10^{15}$ cm^{-2}	Depth, nm	Thickness, nm	α
12	235	283	0.85
10	250	249	0.86
8	265	218	0.92
6	295	160	0.7
4	320	106	0.71

The geometry of the experiment is shown in Figure 1. A femtosecond excitation pulse of the second harmonic of a Ti:sapphire laser was focused on a spot of 3 μm diameter on the surface of the sample. The parameters of the pulse were: τ= 160 fs, λ= 400 нм, E= 0.4 nJ. The minor ($10^{-6} \div 10^{-3}$) changes in the amplitude and in the phase of the first-harmonic pulse of the same laser beam (λ= 800 nm) which was reflected by the sample under investigation, were detected with a Sagnac interferometer [5]. In this interferometer type, the laser pulse is split into two beams, the reference beam and the probe beam. The pulse from the reference beam reaches the specimen before and the pulse from the probe beam after the excitation pulse. The interference of the reference and the probe pulses at the detector generated signals which were proportional to the phase difference $\delta\varphi = \varphi(\tau)-\varphi(\tau-\Delta T)$, as well as to the sum of relative changes in the modulus $\delta R(\tau)/R+\delta R(\tau-\Delta T)/R$ of the reflection coefficient at times τ and τ-ΔT. The sensitivity was ~1 μrad for the phase and ~10^{-6} for the amplitude of the reflection coefficient, respectively; the spatial resolution was ~2 μm, and the time resolution was <1 ps. To study the spatiotemporal distribution of the elastic wave field, the surface of the specimen was scanned with the exciting beam. The experimental setup made it possible to change the time delay τ between the exciting and probing pulses within an interval of 0 to 4 ns.

3. Discussion

In the structures studied, SAW is excited due to the thermoelastic effect after the absorption of the short laser pulse. The experimental data shown below were obtained by registration of the phase changes of the reflection coefficient of the studied layers under propagation of the elastic pulse. We suggest that the variation in the phase of the reflection coefficient occurs mainly due to the displacement of the surface of the specimen and not due to the photoelastic effect.

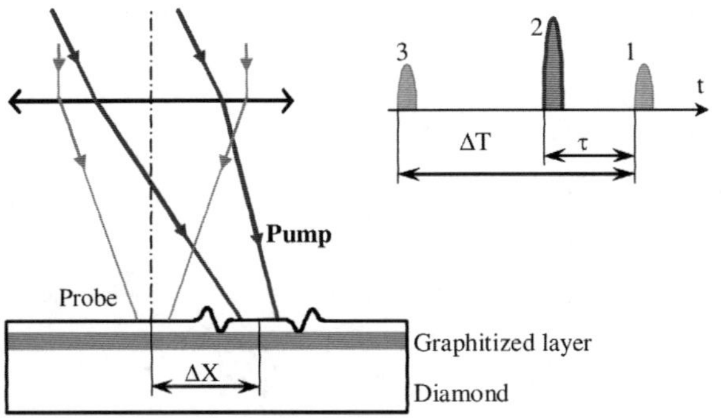

Figure 1. Schematic experimental setup: probe pulse (1) delayed relative to the pump pulse (2). The delay time τ may vary. ΔT is the interval between the arrival of the reference (3) and probe (1) pulses at the sample.

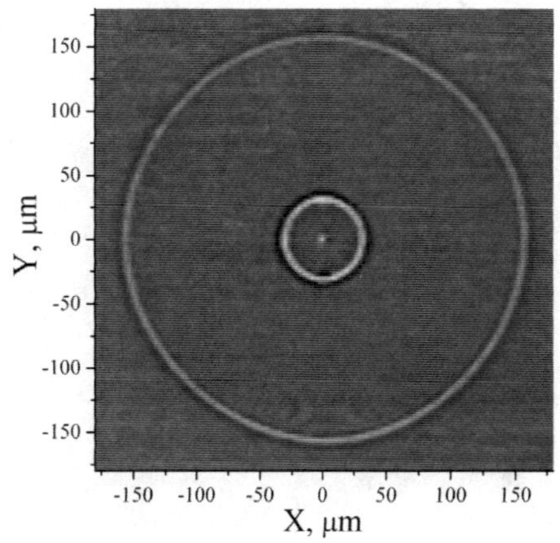

Figure 2. Image of SAW wavefronts. Probe pulse delayed relative to the pump by 3.8 ns

Figure 2 shows the pattern of the SAW field for the area with the implantation dose of $4{\cdot}10^{15}$ cm^{-2}. The probing pulse was delayed by 3.8 ns relative to the exciting pulse. The field distribution appears as concentric rings corresponding to consecutive excitation pulses repeated at 76 MHz frequency. Because of rather low elastic anisotropy of diamond, spatial anisotropy of the SAW propagation was practically absent. A similar pattern of SAW was registered also for area with the implantation dose of $12{\cdot}10^{15}$ cm^{-2}.

To estimate the frequency dispersion, the detection of SAW pulses was performed at various distances between the exciting and the probing spots. Figures 3a and 3b show the shapes of the SAW pulses detected for areas with the doses of $12{\cdot}10^{15}$ cm^{-2} and $4{\cdot}10^{15}$ cm^{-2} respectively.

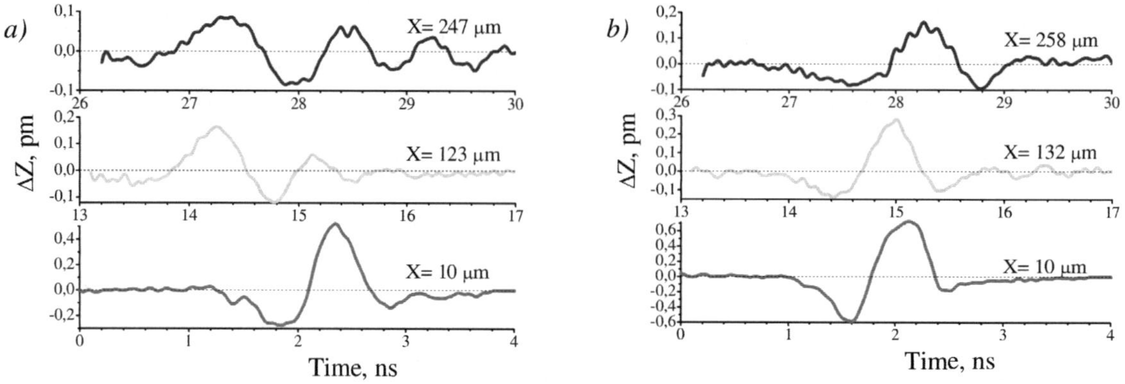

Figure 3. SAW pulses detected *a)* dose $12{\cdot}10^{15}$ cm^{-2}; *b)* dose $4{\cdot}10^{15}$ cm^{-2}. (X - the distance between pump and probe spots)

In both cases, at short distances (~10 μm) between the points of excitation and probing the duration of the pulse was about ~1.2 ns. However, when the distance increased to ~130 μm and further to ~250 μm, the shape of the SAW pulse changed significantly for the area with the higher implantation dose, showing an extension and oscillations. At the same time, for the area with the implantation dose of $4{\cdot}10^{15}$ cm^{-2}, the shape of the pulse showed only a moderate broadening with the increase of the distance.

We performed spectral analysis of the SAW pulses detected at various distances between the exciting and the probing spots. At distances less than 130 μm the spectral width of a SAW pulse was about ~2 GHz, with a maximum at ~700 MHz for all studied areas. For example Figures 4a and 4b shows power spectrum of SAW pulses shown at Figures 3a and 3b.

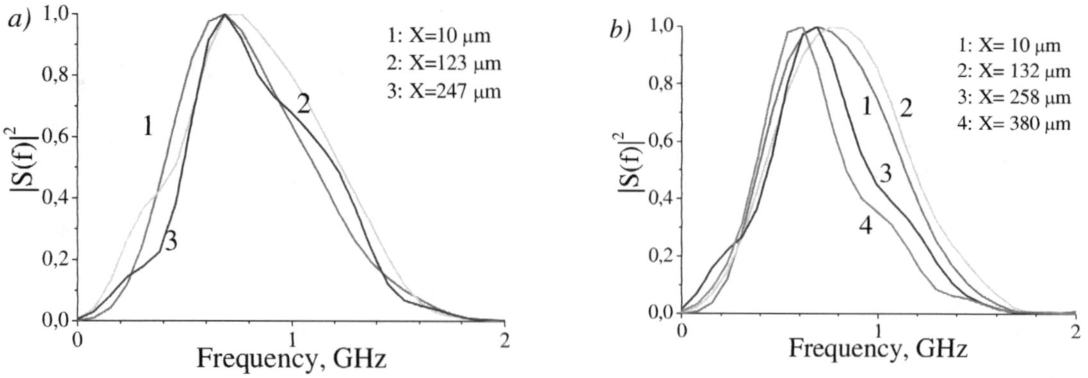

Figure 4. Power spectrum of SAW pulses detected (X - the distance between pump and probe spots). *a)* dose $12{\cdot}10^{15}$ cm^{-2}; *b)* dose $4{\cdot}10^{15}$ cm^{-2}.

With the increase of the distance the pulse spectra for the areas with the implantation doses of $4\cdot10^{15}$ cm^{-2} and $6\cdot10^{15}$ cm^{-2} narrowed to ~1.5 GHz due to an increased attenuation of the high frequency components in course of propagation of the elastic pulse (Figure 4b). This explains the observed moderate broadening of the SAW pulse for area with dose $4\cdot10^{15}$ cm^{-2} at a distance of ~260 μm (Figure 3b). For the areas with the doses of 8, 10 and $12\cdot10^{15}$ cm^{-2} the shape of the spectrum was essentially independent of the distance between the exciting and the probing spots (see for example Figure 4a). Therefore, a significant change in the shape of the pulse (Figure 3a) in course of its propagation can be explained by frequency dispersion.

We estimated the phase velocity of SAW propagation over the range of areas of the studied structure similarly to [8]. For the area with a dose of $4\cdot10^{15}$ cm^{-2} the velocity did not depend on frequency in a range of up to ~2 GHz and was equal to ~9700 m/s, while for the area with a dose of $12\cdot10^{15}$ cm^{-2} the velocity changed from 9900 m/s to 9400 m/s with the increase of frequency by up to 2 GHz. Remarkably, these values are close to the velocity of SAW on intact diamond (~10500 m/s).

In conclusion, we succeeded in detection of SAW excited by subpicosecond optical pulses in diamond with built-in graphitized layer. The absence of anisotropy and considerable difference of acoustical properties from that of diamond makes such layers useful for forming various reflecting, slowing and wave guiding structures.

4. Acknowledgement

We are grateful to V.A.Dravin for diamond implantation, V.P.Martovitsky for X-ray diagnostics and fruitful discussion, and T.I.Galkina for her interest and remarks. This study was supported by the Russian Foundation for Basic Research (projects nos 11-02-01476, 12-02-31865) and by the Russian Academy of Sciences Program of “Basics of Fundamental Research of Nanotechnologies and Nanomaterials”.

References

[1] Krauss A R, Auciello O, Gruen D M, Jayatissa A, Sumant A, Tucek J, Mancini D C, Moldovan N, Erdemir A, Ersoy D, Gardos M N, Busmann H G, Meyer E M and Ding M Q 2001 *Diamond and Related Materials* **10**, 1952-61

[2] Gippius A A, Khmelnitsky R A, Dravin V A and Tkachenko S D *Diamond and Related Materials* 1999 **8** 1631-34

[3] Klokov A, Kochiev M, Sharkov A, Tsvetkov V and Khmelnitskiy R 2011 Graphitized layer buried in a diamond: photothermal properties and hypersound generation under picosecond optical excitation *J. Phys.: Conf. Ser.* **278** 012019

[4] to be published

[5] Tachizaki T, Muroya T, Matsuda O, Sugawara Y, Hurley D H and Wright O B 2006 *Rev. Sci. Instr.* 77, 043713

[6] Khmelnitskiy R A, Dravin V A and Gippius A A 1996 *J. Chem. Vapor Deposition* **5** 121-5

[7] Khomich A V, Khmelnitskiy R A, Dravin V A, Gippius A A, Zavedeev E V and Vlasov I I 2007 *Physics of the Solid State* **49**(9) 1661-65

[8] Neubrand A and Hess P 1992 *Journal of Applied Physic* **71**(1) 227-238

Journal of Physics: Conference Series **520** (2014) 012007 doi:10.1088/1742-6596/520/1/012007

Numerical Simulation of Nonlinear Ultrasonic Waves Due to Bi-material Interface Contact

S Hirose[1] and T Saitoh[2]

[1]Department of Mechanical & Environmental Informatics, Tokyo Institute of Technology, Japan
[2]Division of Environmental Engineering Science, Gunma University, Japan

E-mail: shirose@cv.titech.ac.jp

Abstract. Boundary integral equations are formulated to investigate nonlinear waves generated by a debonding interface of bi-material subjected to an incident plane wave. For the numerical simulation, the IRK (Implicit Runge-Kutta method) based CQ-BEM (Convolution Quadrature-Boundary Element Method) is developed. The interface conditions for a debonding area, consisting of three phases of separation, stick, and slip, are developed for the simulation of nonlinear ultrasonic waves. Numerical results are obtained and discussed for normal incidence of a plane longitudinal wave onto the nonlinear interface with a static compressive stress.

1. Introduction

Nonlinear ultrasonic nondestructive testing has been developed over the last decade [1], since nonlinear waves are generated by nonlinear elasticity of materials and unbonded interface conditions and are very sensitive to a degradation of material properties at very early stage. However, the mechanism of generation of nonlinear waves has not yet been understood very well from the theoretical and/or numerical point of view.

So far, two dimensional simulations on nonlinear ultrasonic waves have been carried out [2][3]. Also, the axisymmetric problem of a circular crack subjected to normal incidence of a longitudinal wave was solved numerically [4]. However, no full three dimensional analysis has been done. It is, therefore, demanded to conduct numerical simulations with three dimensional realistic models. In this paper, three dimensional boundary integral equations are formulated for a circular interface crack with unbonded conditions in bi-material half spaces. The integral equations are discretized using the IRK based CQ-BEM and numerically solved to investigate the nonlinearity involved in ultrasonic waves scattered by the nonlinear interface crack.

2. Formulation of boundary integral equations

The model of a bi-material, consisting of two semi-infinite domains D and D^{II}, is shown in figure 1. S_h and S_d represent the bonding and debonding areas, respectively, of the bi-material interface. Assuming that an incident plane wave is given in D^I, we may formulate time-domain boundary integral equations, in which reflected and transmitted waves are unknown values. However, the time-domain boundary integral equations for the bi-material have the following disadvantages:

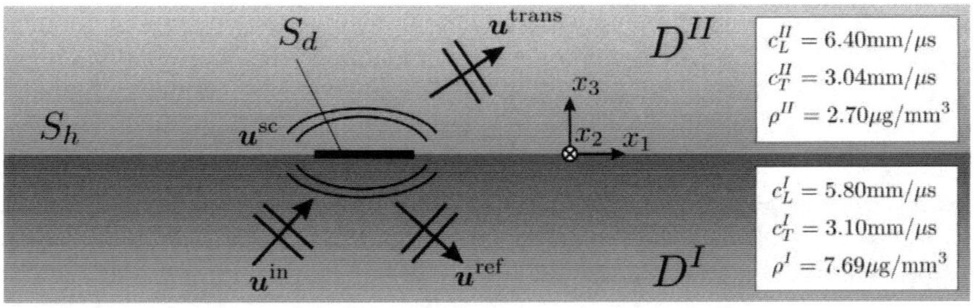

Figure 1. Model for numerical simulation.

- For normal incidence of a plane wave, reflection and transmission occur at all points on the interface from the initial step in time. Due to this fact, truncation errors can be introduced at the edge elements, where the infinite interface is truncated in numerical analysis.
- For oblique incidence, initial conditions that are given by zero displacement and velocity for unknown reflected and transmitted waves at all points in the domains, cannot be satisfied unless an infinite interface is taken into account in the numerical analysis.

Therefore, in this paper, the integral formulation in which unknown variables are only scattered waves from the debonding area, is proposed in order to overcome these difficulties. If the flat interface of infinite extent is perfectly bonded and is subjected to a plane wave incidence, it is easy to calculate analytically the "free field $\boldsymbol{u}^{\text{free}}$", defined as the summation of the incident wave $\boldsymbol{u}^{\text{in};I}$ and the reflected wave $\boldsymbol{u}^{\text{ref};I}$ in D^I, and the transmitted wave field $\boldsymbol{u}^{\text{trans};II}$ in D^{II} as follows:

$$\boldsymbol{u}^{\text{free};I} = \boldsymbol{u}^{\text{in};I} + \boldsymbol{u}^{\text{ref};I} : D^I, \qquad \boldsymbol{u}^{\text{free};II} = \boldsymbol{u}^{\text{trans};II} : D^{II} \tag{1}$$

If debonding exists in a local area on the interface, the free field may be disturbed by the wave field $\boldsymbol{u}^{\text{sc}}$ scattered by the debonding area, and the total displacement $\boldsymbol{u}$ can be expressed by $\boldsymbol{u} = \boldsymbol{u}^{\text{sc}} + \boldsymbol{u}^{\text{free}}$. Since the scattered wave $\boldsymbol{u}^{\text{sc}}$ satisfies the initial condition and the radiation condition, the boundary integral equations for $\boldsymbol{u}^{\text{sc}}$ can be formulated. In solving the boundary integral equations for $\boldsymbol{u}^{\text{sc}}$, the convolution integrals with time are evaluated by IRK based CQM [5] and the surface integrals over the bonding and debonding interfaces are discretized by constant elements. For the acceleration, the fast multipole method is applied to IRK based CQ-BEM [6].

The boundary integral equations for $\boldsymbol{u}^{\text{sc}}$ in D^I and D^{II} are simultaneously solved using appropriate interface conditions. The interface condition on the bonding area S_h is the continuity of displacement and stress. For the debonding area S_d, three types of interface conditions, "separation", "stick", and "slip", are considered [2]. "separation" means that two surfaces of upper and lower materials are separated with no traction, while "stick" and "slip" are contact conditions under compressive normal stress state. For the "stick" condition, the surfaces of two materials move with no relative velocity. On the other hand, the "slip" condition allows a relative tangential movement with dynamic friction force.

3. Numerical examples

Numerical examples are presented for nonlinear ultrasonic wave problems of bi-material interface subjected to a small static compressive stress 0.74kPa normal to the interface and the normal incidence disturbance of longitudinal waves with 2 and 4MHz frequencies. The incident wave is a sinusoidal plane wave with three cycles and 10nm amplitude. The debonding area is a circular interface crack with radius 0.5mm. We assume that the material constants for bi-material are given in figure 1. The coefficients of static and kinematic friction are given by 0.61 and 0.47, respectively.

3.1. Vertical displacement at the center points on top and bottom debonding area

In figures 2 (a) and (b), the vertical displacements at the center points on the top and bottom debonding surfaces, subjected to 2MHz and 4MHz sinusoidal incident waves, respectively, are demonstrated as a function of time. Note that the time scales in these two figures are different. The bottom surface is forced to move according to the direct incidence of the sinusoidal wave, whereas the top surface is freely movable for a short time after deformation together with the bottom surface in contact condition, and then affected by the diffracted waves from the edge of the circular debonding area. Thus the periods of vibrations of top and bottom surfaces are different, and the dynamic behaviors of the debonding surfaces, especially the top surface, are different depending on the frequency. Note that these clapping motions of the debonding interface in three dimensional problem are very similar to the results of two dimensional simulations [3].

3.2. Vertical displacement 2.0mm above the center of debonding area

Figure 3 shows the vertical displacements of total and scattered wave fields ((a) and (b)) at the internal point located 2.0mm above the center of the debonding area, and the normalized frequency spectra of the scattered waves $\boldsymbol{u}^{sc}$ ((c) and (d)). Figures (a) and (c) are the results for the 2MHz sinusoidal incident wave, and figures (b) and (d) are for the 4MHz case. Note that the vertical axes for the vertical displacements of scattered waves are shown on the right side of the graphs. For comparison, the results for the displacements of the free fields in the case of no debonding are shown.

In both figures of (a) and (b), the time variations of vertical displacements of total waves show periodic waveforms and small aftereffects. However, the aftereffects do not continue for a long time compared with the results of two dimensional simulations [3]. In the case of 2MHz, the vertical displacement of scattered wave generated by the clapping motion shows a shorter period than the fundamental frequency of 2MHz, and hence large second and higher harmonics as well as sub-harmonics are seen in the frequency spectrum (see figure (c)). In the case of 4MHz, on the other hand, the waveform of scattered wave is a little distorted compared with the case of 2MHz, and the frequency spectrum shows relatively small amplitudes of the second and third harmonics. From these results, it can be said that the generation of nonlinear ultrasonic waves like higher harmonics and sub-harmonics due to contact conditions on the debonding interface depend largely on the frequency.

4. Conclusions

In this paper, the boundary integral equations for bi-material subjected to an incident plane wave were formulated, and the IRK based CQ-BEM was implemented in the numerical simulation. The interface conditions for debonding areas that consists three phases of "separation", "stick", and "slip", were developed for the simulation of nonlinear ultrasonic waves. Numerical results showed that the generation of nonlinear ultrasonic waves are largely dependent on the frequency as well as the contact conditions on the interface.

References

[1] Solodov I, Doring D, and Busse G 2011 *J. Mech. Eng.* **57** 169
[2] Hirose S 1994 *Wave Motion* **19** 37
[3] Saitoh T, Furuta Y, Hirose S, and Nakahata K 2011 *Journal of Japan Society of Civil Engineers, Ser. A2 (JSCE J. Applied Mech.)* **67** I_161 (in Japanese)
[4] Hirose S and Achenbach J D 1993 *J. Acoust. Soc. Am.* **93** 142
[5] Maruyama T, Saitoh T, and Hirose S 2012 *Proc. ICOM/JASCOME 2012* **12** 91 (in Japanese)
[6] Saitoh T and Hirose S 2010 *WCCM2010, IOP Conf. Series: Materials Science and Engineering* **10** 012242

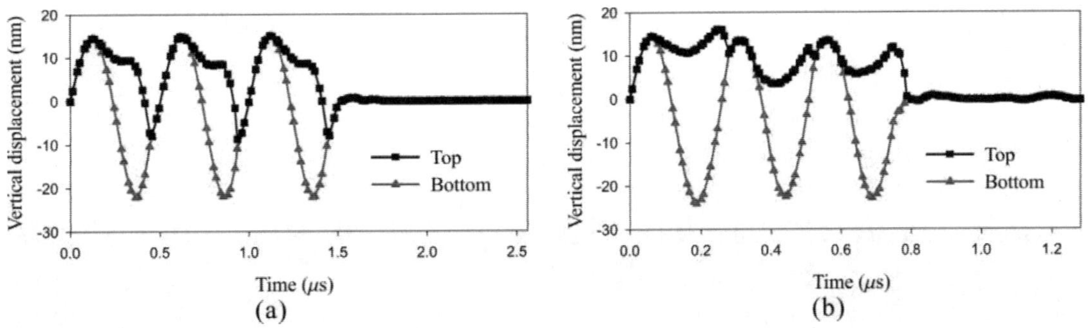

Figure 2. Time variations of vertical displacements at the center points on top and bottom debonding surfaces in the cases of normal incidence of sinusoidal longitudinal waves with (a) 2MHz and (b) 4MHz.

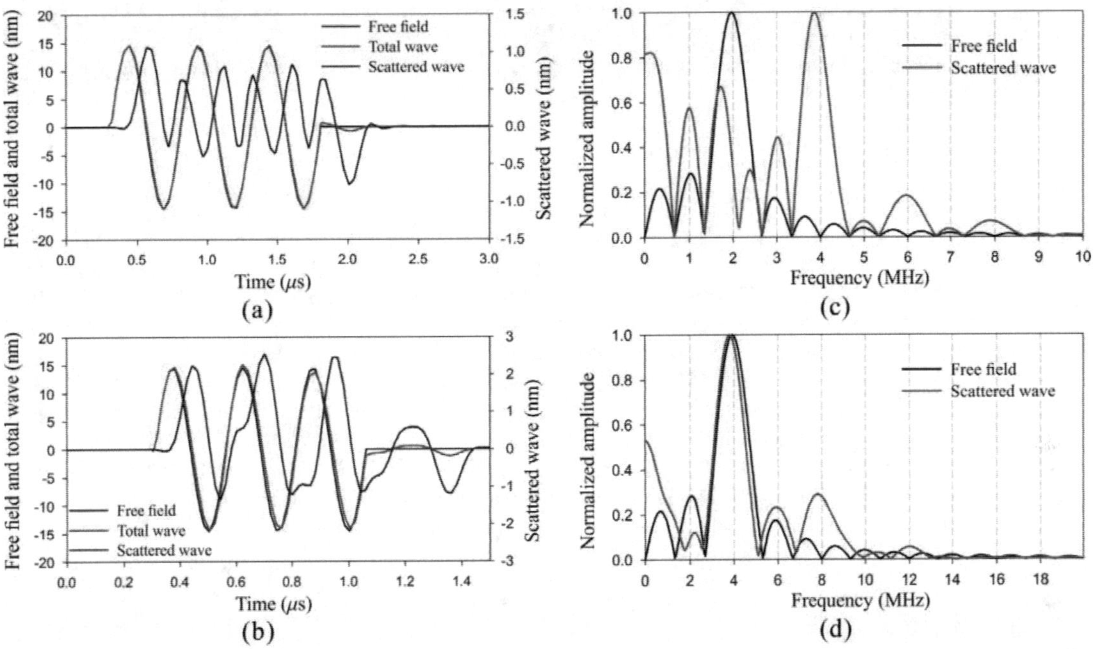

Figure 3. Vertical displacements as a function of time ((a) and (b)) and frequency spectra of scattered waves ((c) and (d)) at the point 2.0mm above the center on top debonding surfaces in the cases of normal incidence of sinusoidal longitudinal waves with 2MHz (top) and 4MHz (bottom) frequencies.

Journal of Physics: Conference Series **520** (2014) 012008 doi:10.1088/1742-6596/520/1/012008

Dynamic Material Property Measurement of Steel Thin Sheets using Laser-Based Ultrasonics

Y Nagata[1], H Yamada[1], S Hashiguchi[1], C S Lim[2], H C Park[2], H J Huh[2], M K Kang[2] and K J Oh[2]

[1] Process Research Laboratories, Nippon Steel & Sumitomo Metal Corp., 20-1 Futtu-shi, Chiba, 293-8511 Japan
[2]POSCO Technical Research Laboratories ,1, Goedong-dong, Nam-gu, Pohang, 790-785, Republic of Korea

E-mail: nagata.93x.yasuaki@jp.nssmc.com

Abstract. A material property measurement system for steel sheets using laser-based ultrasonics was developed. The system consists of a pulsed Nd:YAG laser for ultrasonic generation and multi-channel interferometer coupled with a CW single frequency laser for ultrasonic detection. The system can measure the frequency of the S1 Lamb wave mode of zero group velocity (S1f) as well as the longitudinal resonance frequencies without ablative damage to the steel surface. It was confirmed that Poisson's ratio could be directly obtained by combining the measured S1f value and the longitudinal resonance frequencies. To evaluate the applicability of this system in an actual steel production setting, the system was installed in hot rolling pilot plant that produces steel samples. As a result, it was demonstrated that the system can measure dynamic changes in Poisson's ratio values within steel sheets, even in the hot rolling pilot plant environment. Material property data, such as Poisson's ratio, during the thin sheet production process will be very useful for manufacturing high value-added steel, such as sheets with uniform quality.

1. Introduction

Laser-based ultrasonic technology, where the ultrasonic waves are generated and detected by lasers, is a promising technology for non-destructive and non-contact evaluations of the material properties of moving and/or high temperature targets. Especially in the field of steel production, laser-ultrasonic technology is considered to be a very attractive solution for in-process measurements of the properties of the steel on the production line. Even though fundamental technologies have already been developed and have resulted in huge progress in application technologies, several improvements are still needed to be made in order to realize in-process measurements of the microstructure of steel materials in an actual production line setting.

Material property data, such as Poisson's ratio, during the thin sheet production process would be very useful for manufacturing high value-added steel, such as sheets with uniform quality. However, until now, no in-process system for measuring Poisson's ratio in steel sheets existed. In this study, the development of a system using laser ultrasonics to measure Poisson's ratio was explored. In particular, ultrasonic generation techniques in the thermoelastic region were developed in order to avoid ablative damages to the surface of the sheets.

Furthermore, the robustness of the measurement system in an actual production line setting was explored. To confirm the robustness, a prototype measurement system was installed in a hot rolling pilot plant, and its performance was tested.

2. Preliminary experiments to detect the Lamb waves of zero group velocity

In order to apply laser ultrasonics to steel samples without causing surface damage, the energy of the pulsed Nd:YAG laser used for ultrasonic generation should be controlled at a relatively low level, and the generation scheme should be in the thermoelastic region rather than the ablation region. Clorennec D. et al. reported on using pulsed laser generation in the thermoelastic region for efficiently generating first-order symmetric Lamb mode (S1) in Duralumin plates under laboratory conditions [1]. It was reported that this S1 mode exhibited an anomalous behavior at the frequency (S1f) in which the group velocity vanished.

At first, preliminary experiments and analyses were conducted in order to investigate whether S1f measurements were possible for steel sheets as well as whether such measurements were effective for evaluating the material properties. A Q-switched Nd:YAG laser of the wavelength of 1064 nm was used for the ultrasonic generation, and 1W CW single frequency laser of the wavelength of 532 nm was used for the ultrasonic detection. Both of ultrasonic generation and detection laser beams were adjusted to the same spot on the sample surface. In this experimental system, a multi-channel interferometer was adopted [2]. According to our experiments, the interferometer showed better robustness about a liftoff deviation, compared with a conventional Fabry-Pérot interferometer combined with the 1W detection laser. Therefore, the multi-channel interferometer was adopted (liftoff: distance between sample surface and the front part the probe head).

The experiments were conducted for a 2 mm thick plain carbon steel sheet at room temperature, and the fluence of the ultrasonic generation laser beam was controlled at 1.8 mJ/mm^2, which is in the thermoelastic region. Figure 1 shows a FFT spectrum of the detected waveform, and a spectrum peak at 1.3 MHz was clearly observed. Figure 2 shows the theoretically calculated relationship between the frequency and the group velocity of Lamb waves in the case of a 2 mm thick steel sheet. In the calculations in figure 2, the longitudinal and shear wave velocity values were assumed to be the typical values of 5900 m/s and 3200 m/s, respectively. According to figure 1 and figure 2, it was confirmed that the spectrum peak in figure 2 corresponded to the S1 Lamb wave mode of zero group velocity (S1f).

It was reported that S1f can be expressed theoretically as follows, where V_L is longitudinal wave velocity, d is sample thickness and β1 is a function related to Poisson's ratio [3]. It was additionally reported that β1 values can be deduced from the dispersion curves of Lamb waves. Figure 3 shows the theoretical relationship between the β1 coefficient and Poisson's ratio, and it shows that β1 varies from 0.975 to 0.60 as Poisson's ratio varies from 0 to 0.451 [3].

$$\mathrm{S1f} = \beta 1 \cdot \frac{V_L}{2d} \tag{1}$$

Since Poisson's ratio is related to longitudinal and shear wave velocities, S1f is also related to longitudinal and shear wave velocities. If S1f values can be obtained dynamically, it will be possible to monitor the dynamic changes in material characteristics.

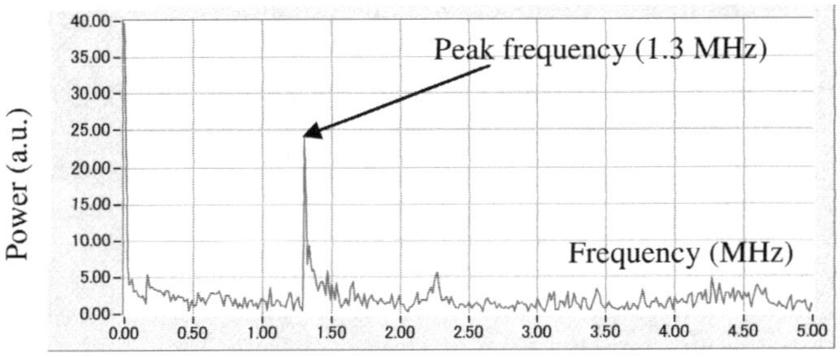

Figure 1. FFT spectrum of the detected waveform.
(FFT was conducted for a 90 μs temporal period of the waveform after the generation timing.)

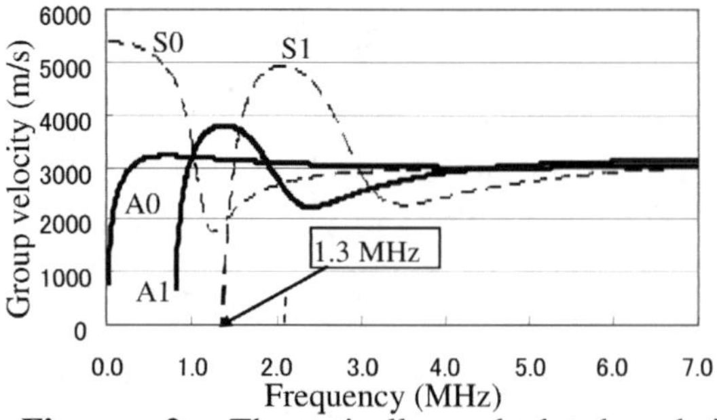

Figure 2. Theoretically calculated relationship between the frequency and the group velocity of Lamb waves in the case of 2 mm thick steel sheets.

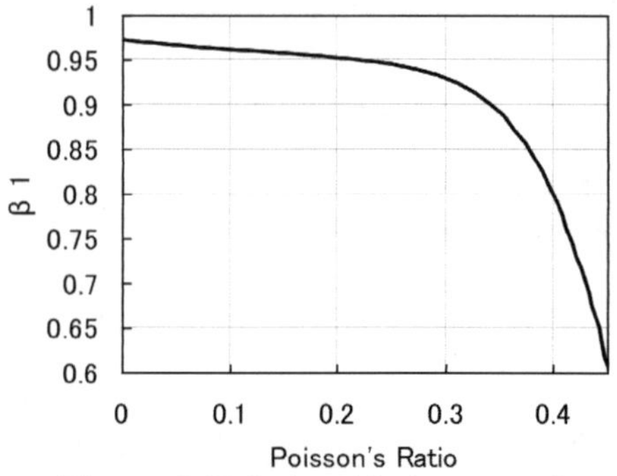

Figure 3. Relationship between the β1 coefficient and Poisson's ratio.

3. Development of a technique to measure Poisson's ratio

We have developed a system for measuring S1f values in the thermoelastic region, and the developed system was installed in hot rolling pilot plant. Figure 4 shows the schematic diagram of the pilot plant trial. Samples were heated in a furnace, and after descaling by the descaling machine, they were then hot rolled several times and ultrasonic measurements were conducted. The spot diameter of the focused generation beam was about 4 mm, and the power density at the irradiation spot was kept low in order to generate ultrasonics in thermoelastic region (fluence: 2.0 mJ/mm^2). A compact head component for ultrasonic generation and detection was developed for the pilot plant trials.

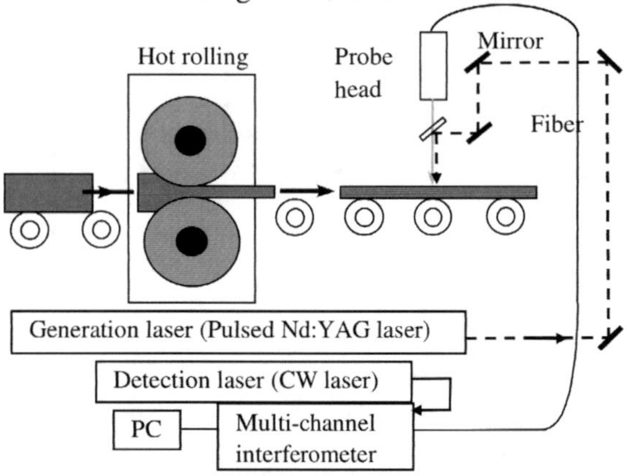

Figure 4. Prototype measurement system installed in a hot rolling pilot plant. (Special large diameter collection lens of the scattered light from the sample surface was made and adopted in the probe head part. Its diameter was 3 inches, rather than standard 2 inches.)

Figure 5 shows an example of FFT analysis for a typical waveform. Compared to the outcome shown in figure 1, a highly sensitized detection section allows the observation of plural frequency peaks, as shown in figure 5. The space between the respective frequency peaks, which resulted from the investigation to identify each frequency peak, is indicated in figure 6. It emerged from these results that while the frequency peak of the lowest peak corresponds to S1f, the other frequency peaks correspond to the resonant frequencies of the longitudinal waves since the space between each of the frequencies are all equal. The space between the resonant frequencies of the longitudinal waves Δf is theoretically calculated by $\Delta f = V_L/(2d)$. Consequently, it also emerged that β 1 can be calculated with the S1f and Δf values using equation (1), and Poisson's ratio can thus be obtained using the data in figure 3. Figure 7 shows the results of consecutive Poisson's ratio measurements using a measurement method at high temperatures, including temperatures in the vicinity of phase transformation as the steel sample naturally air-cooled. The above results enabled us to confirm that it is possible to dynamically measure Poisson's ratio at high temperatures. In order to evaluate the validity of the developed technique, five steel samples of different steel kind were prepared and each Poisson's ratio was measured by the developed technique at room temperature. Additionally, longitudinal and shear wave velocity of each sample were measured by using contact probes, and each Poisson's ratio was calculated by the following theoretical equation, where V_S is shear wave velocity.

$$\text{Poisson's ratio } = \frac{1}{2}\left\{1 - \frac{1}{(V_L / V_S)^2 - 1}\right\} \tag{2}$$

Figure 8 is the comparison of Poisson's ratio measured by contact probes and developed laser ultrasonic technique. It was confirmed that Poisson's ratio measured by laser ultrasonic technique showed a good agreement with the values measured by contact probes.

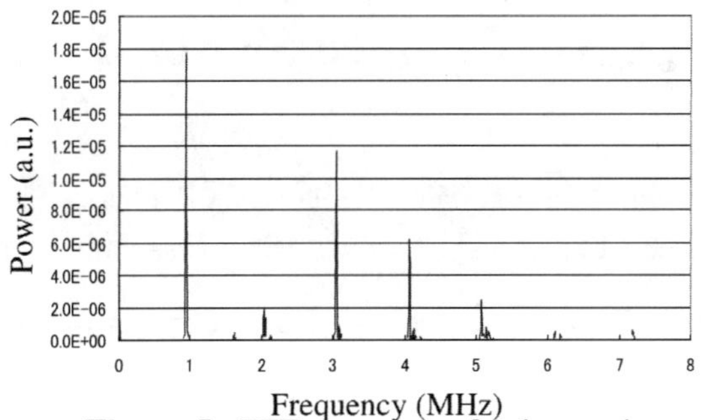

Figure 5. FFT spectrum of a detected typical waveform.

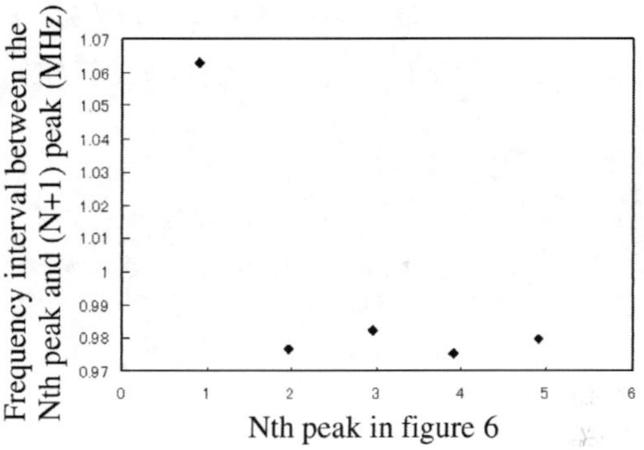

Figure 6. Characteristic of the frequency peaks in figure 5.

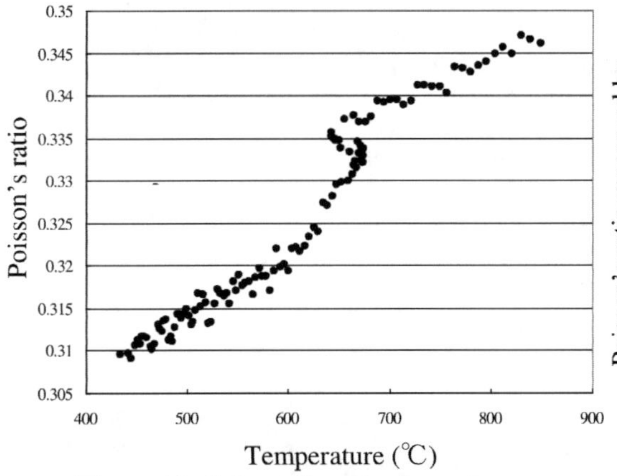

Figure 7. Poisson's ratio dependence on sample temperature. (Sample: High carbon steel, Thickness: 3 mm)

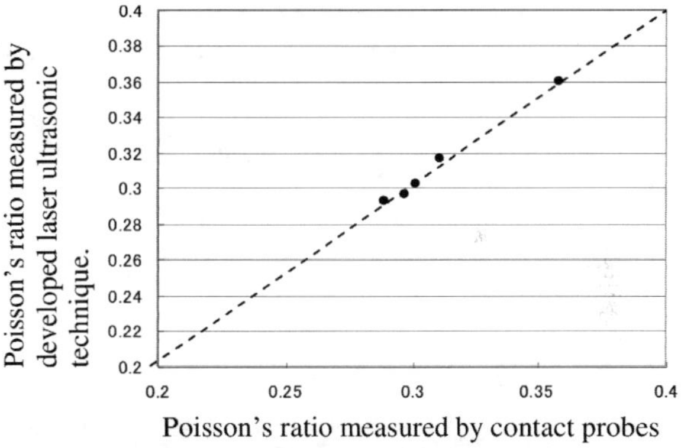

Figure 8. Comparison of Poisson's ratio measured by contact probes and developed laser ultrasonic technique.

4. Conclusions

The developed system is capable of measuring Poisson's ratio in steel sheets using the frequency data of S1 Lamb wave mode of zero group velocity (S1f) and longitudinal wave resonance frequencies without ablative damages on the steel surface, even in a hot rolling pilot plant environment.

References

[1] Clorennec D, Prada C, Royer D and Murray T W 2006 *Appl. Phys. Lett.* **89** 024101
[2] Pouet B, Breugnot S and Clémenceau 2006 *Review of Progress in Quantitative Nondestructive Evaluation* **25A and 25B**, ed. by D O Thompson and D E Chimenti (New York: American Institute of Physics) 233
[3] Clorennec D, Prada C and Royer D 2007 *J. Appl. Phys.* **101** 034908

Journal of Physics: Conference Series **520** (2014) 012009 doi:10.1088/1742-6596/520/1/012009

Flaw Inspection of Aluminum Pipes by Non-Contact Visualization of Circumferential Guided Waves using Laser Ultrasound Generation and an Air-Coupled Sensor

K Urabe[1], J Takatsubo[2], N Toyama[1], T Yamamoto[1] and H Tsuda[1]

[1] National Institute of Advanced Science and Technology, 1-1-1 Umezono, Tsukuba 305-8568, Japan
[2] Tsukuba Technology Co., Ltd., Plaza214, 2-1-6 Sengen, Tsukuba 305-0047, Japan

E-mail: urabe-k@aist.go.jp

Abstract. Our group had previously proposed a generation laser scanning system for visualizing ultrasound propagation on an object as an animate image, which provided visible and quick flaw inspection. Recently, we improved this system to make it completely non-contact by employing an air-coupled ultrasound transducer as a receiver instead of a contact transducer, and demonstrated the successful visualization of Lamb waves propagating on aluminum and carbon fiber reinforced plastic plates, as well as the detection of flaws. In this research, we applied this system to the non-contact visualization of circumferential guided waves on aluminum pipes. It was shown that circumferential guided waves propagating in opposite directions could be visualized separately, and that a flaw such as a slit or thinning on the inside surface of the pipe could be successfully detected even when it existed outside the scanning area.

1. Introduction

Our group had previously proposed a generation laser scanning system for visualizing ultrasound propagation on an object as an animate image, for the purpose of realizing "simple," "quick," "easy to understand," and "non-contact" ultrasound non-destructive inspections [1][2]. Recently, we improved this system to make it completely non-contact by employing an air-coupled ultrasound transducer as a receiver, instead of the conventionally used contact piezoelectric transducer [3][4]. Using this non-contact ultrasound visualization system, we demonstrated the successful visualization of Lamb waves propagating on plate-shaped specimens. From the visualized images, flaws such as delamination or lack of fiber in carbon fiber reinforced plastics and slits or thinning on the backside of aluminum plates could be detected.

In this research, we investigated an application of this system to the non-contact visualization of circumferential guided waves, which are guided waves propagating in the circumferential direction, on pipes, with the aim of inspecting for flaws on the inner side of the pipes. The possibility of inspecting the entire circumference is an advantage of the non-destructive inspection of a pipe using circumferential guided waves [5]. The possibility of using a higher frequency compared to guided waves propagating in the axial direction, which would provide a higher sensitivity to small defects, has also been pointed out as an advantage [5]. Several studies have been conducted on the non-contact generation and detection of circumferential guided waves and improvements in the received signal amplitude, with the purpose of non-contact inspections for the thinning of pipes [6][7].

2. Experimental setup

The experimental setup is shown in figure 1. A pulsed laser beam was irradiated from the side of the pipe, and a wide-band ultrasound signal was generated at the irradiated point of the pipe by thermal strain. From the generated ultrasound, circumferential guided waves were received by the air-coupled transducer, which was placed above the pipe. The irradiated point was scanned with a fixed pitch in the circumferential direction on the pipe using a mirror scanner, and the pipe was moved in the axial direction with a fixed pitch. Thus, a two-dimensional area of the pipe was scanned by the laser beam. The received signal for each irradiated point was amplified, filtered, and recorded by the computer. Subsequently, the received ultrasound signals were rearranged to form a two-dimensional contour map of the received amplitudes corresponding to the irradiated (generated) points, for each time. When showing the maps represented as color or brilliance images continuously in order of time, the sequence appeared to show an animation of the ultrasound propagation, generated by the air-coupled transducer and propagating in the circumferential direction in the scanned area, because of the reversibility of the ultrasound propagation [1][2]. For a pipe specimen, the receiving angle could be adjusted to the critical angle of the guided wave by adjusting the receiving position above the pipe, as shown in figure 2, so as to obtain a high amplitude for the received signal and to select the mode and propagating direction of the wave to be received [6]. The use of a line-focus-type transducer has been proposed to maintain a constant receiving angle for the transducer and thus obtain a higher level for the received signal [7]. To receive the circumferential guided wave at the critical angle, the transducer was better to be flat along the axial direction of the pipe, rather than a point-focus-type one where the receiving angle varies in the axial direction and directivity for the guided wave was lower [3]. Therefore, we used a line-focus-type transducer (Japan Probe Co., Type 0.4 or 0.8K20×20N R90, for 400 or 800 kHz band respectively) with a size of 20 × 20 mm and a 90-mm radius of curvature, and adjusted its position above the pipe (the horizontal and vertical directions shown in figure 2) so as to obtain the maximum amplitude for the received signal. Other experimental conditions are indicated in figure 1, and also in the figure captions in the next section. The gain of the receiver was regulated so that the maximum signal level was ±2.0 V after amplification, and the total gain (dB) is noted in the figure captions of the visualized images. A smaller value for the total gain means a larger level for the received signal. The frequency of the visualized ultrasound was 340 kHz or 700 kHz, where the highest sensitivity was obtained for each transducer. As specimens, aluminum alloy pipes with a diameter of 100 mm, thickness of 5 mm, length of 1 m, and artificial flaws on the inside surface were used.

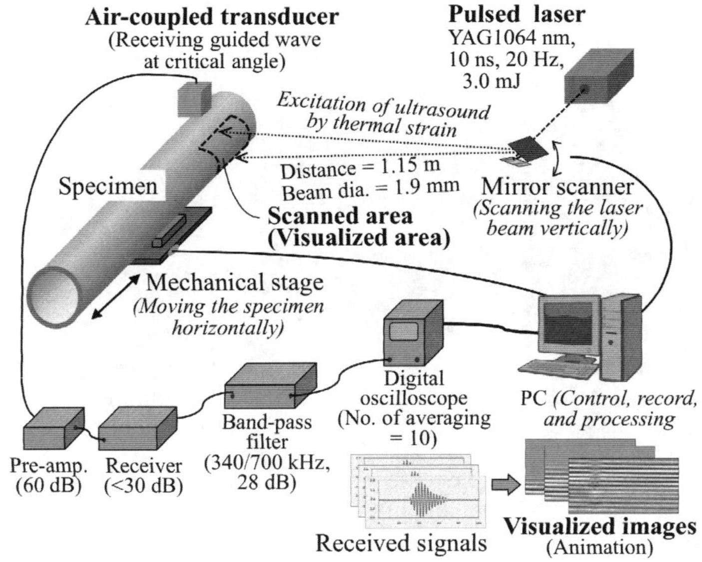

Figure 1. Experimental setup.

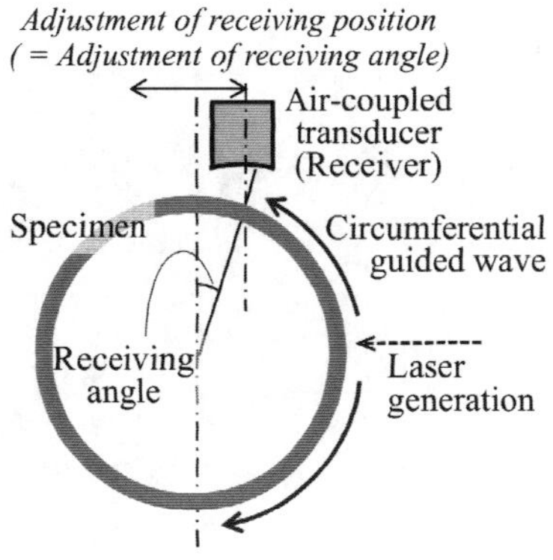

Figure 2. Adjustment of receiving position (receiving angle) and direction of guided wave to be visualized.

Journal of Physics: Conference Series **520** (2014) 012009 doi:10.1088/1742-6596/520/1/012009

3. Results and discussion

Figure 3 shows propagation images of the circumferential guided waves on an aluminum pipe with 3-mm-deep spherical thinning on the inside surface at 340 kHz. The air-coupled transducer was set at 6.0 mm to the right side from just above the pipe axis in figure 3(a), whereas it was set at 6.6 mm to the left side in figure 3(b). Therefore, the propagating directions of the visualized waves were opposite to each other. The time (μs) on each image indicates the time that had passed since the irradiation of the laser beam, which equivalently corresponds, on the image, to the passage of time from the generation of the ultrasound wave at the transducer. As shown in the left and the middle images of figure 3(a), the thinning caused a phase delay when the wave passed it. After some interval, the 2nd round wave appeared as shown in the right image, where the effect of the thinning that was passed in the 1st round appears. Hence, as can be seen in figure 3(b), even if the thinning was not in the scanned (visualized) area, it could be detected as a phase change in the wave in the image, as long as it existed in the propagation area for the circumferential guided wave. The receiving angle values given in each figure caption were derived from the position of the transducer (see figure 2), and they were generally close to the critical angle of Lamb wave of A_0 mode for a 5-mm-thick aluminum plate [3].

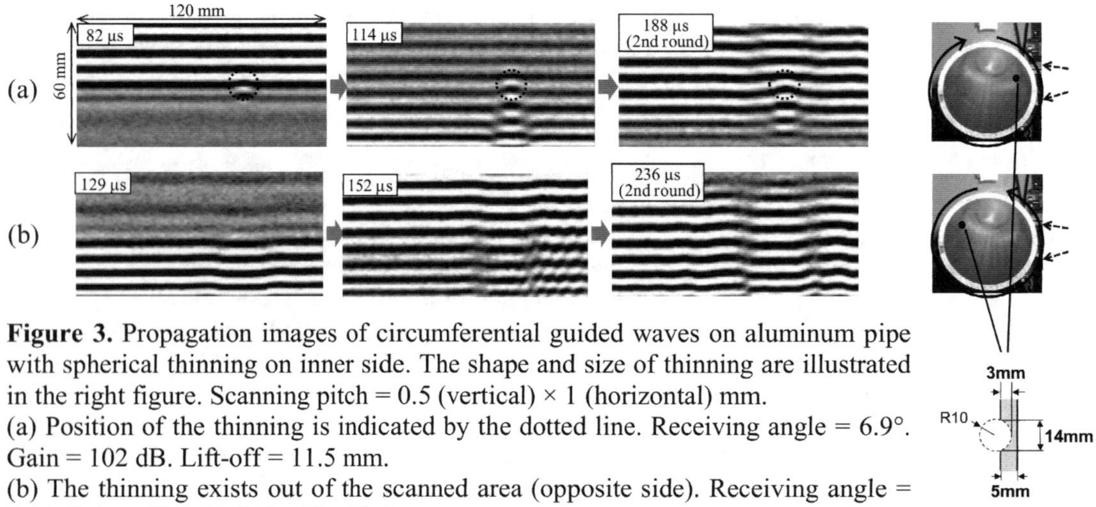

Figure 3. Propagation images of circumferential guided waves on aluminum pipe with spherical thinning on inner side. The shape and size of thinning are illustrated in the right figure. Scanning pitch = 0.5 (vertical) × 1 (horizontal) mm.
(a) Position of the thinning is indicated by the dotted line. Receiving angle = 6.9°. Gain = 102 dB. Lift-off = 11.5 mm.
(b) The thinning exists out of the scanned area (opposite side). Receiving angle = 7.6°. Gain = 106 dB. Lift-off = 13.0 mm.

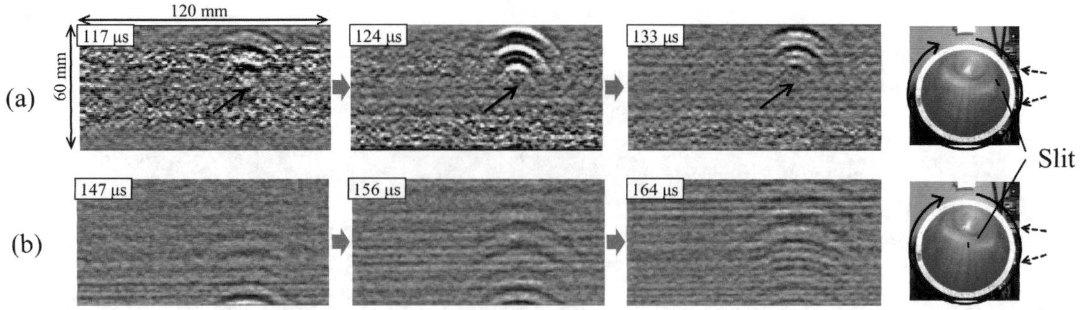

Figure 4. Propagation images, after eliminating forward travelling wave, of circumferential guided waves on an aluminum pipe with slit (4 mm long, 1 mm deep, and 0.3 mm wide) on inner side. Scanning pitch = 0.5 (vertical) × 1 (horizontal) mm.
(a) Position of the slit (center) is indicated by the arrow. Receiving angle = 6.2°. Gain = 96 dB. Lift-off = 21.5 mm.
(b) The slit exists outside the scanned area (bottom). Receiving angle = 6.7°. Gain = 100 dB. Lift-off = 21.0 mm.

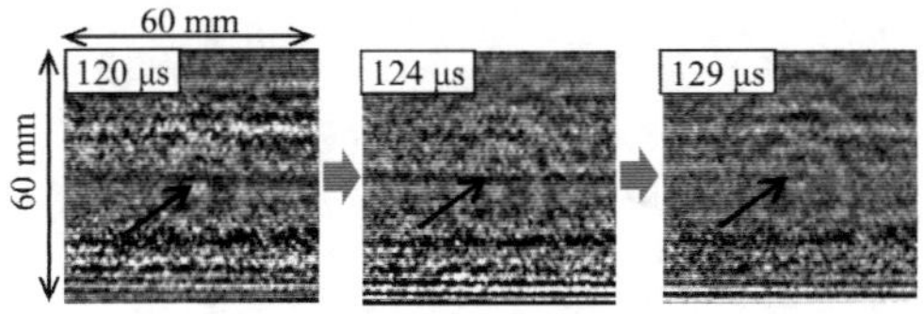

Figure 5. Propagation images when slit (4mm long, 2 mm deep, and 0.3 mm wide) is parallel to propagating direction. The position of the slit is indicated by the arrow. Scanning pitch = 0.5 (vertical) × 0.5 (horizontal) mm.
Receiving angle = 6.7°. Gain = 97 dB. Lift-off = 21.0 mm.

Figure 4 shows images at 700 kHz for a pipe with a 4-mm-long and 1-mm-deep slit on the inside surface. The images are those after eliminating the forward travelling wave using the synchronizing differential method [8], so as to emphasize the reflected or scattered wave caused by the slit. In addition to the slit in the scanned area (figure 4(a)), a slit that existed outside the scanned area (figure 4(b)) was clearly detected. Figure 5 shows propagation images when a 4-mm-long and 2-mm-deep slit was parallel to the propagating direction of the wave. Although it is unclear compared with a slit that was perpendicular to the propagating direction, the wave reflected from the slit was observed.

4. Conclusion

We have shown the non-contact visualization of circumferential guided waves propagating on an aluminum pipe, using generation laser scanning and an air-coupled transducer as a receiver. The propagating direction of the wave to be visualized could easily be selected by adjusting the position of the air-coupled transducer. The propagation of the circumferential wave was successfully visualized as clear images, and flaws such as thinning or a slit on the inside surface of the pipe could easily be detected as a phase change, scattering, or reflection of the propagating wave from the image, even when the flaw was outside the scanned area. The elimination of the forward travelling wave was effective for detecting a slit.

Further research and developments is necessary to apply this system to other materials with various thicknesses and structures with more complicated shapes. Appropriate frequency and mode of the visualized wave, and the conditions for the laser irradiation are important factors that must be further investigated. Studies on the reduction of the measurement time and image processing techniques to emphasize the signals from defects are now in progress.

Acknowledgements

This work was supported by JSPS KAKENHI Grant Number 22560687.

References

[1] Takatsubo J, Wang B, Tsuda H and Toyama N 2007 *Journal of Solid Mechanics and Materials Engineering* **1** 1405

[2] Yashiro S, Takatsubo J, Miyauchi H and Toyama N 2008 *NDT & E International* **41** 137

[3] Urabe K, Takatsubo J, Toyama N, Tsuda H and Nagai H 2012 *Journal of the Japan Society for Composite Materials* **38** 183 (in Japanese)

[4] Urabe K, Takatsubo J, Toyama N, Tsuda H, Wang B and Nagai H 2012 *Journal of the Japanese Society for Non-destructive Inspection* **61** 537 (in Japanese)

[5] Nishino H, Yokoyama R, Kondo H and Yoshida K 2007 *Japanese Journal of Applied Physics* **46** 4568

[6] Nishino H, Asano T, Taniguchi Y, Yoshida K, Ogawa H, Takahashi M and Ogura Y 2011 *Japanese Journal of Applied Physics* **50** 07HC10

[7] Nishino H, Yoshida K, Asano T, Taniguchi Y, Ogawa H, Takahashi M and Ogura Y 2011 *E-Journal of Advanced Maintenance* **2** 181

[8] Takatsubo J, Miyauchi H, Urabe K, Tsuda H, Toyama N and Wang B 2009 *Transactions of the Japan Society of Mechanical Engineers* **A75** 211 (in Japanese)

Journal of Physics: Conference Series **520** (2014) 012010 doi:10.1088/1742-6596/520/1/012010

Improvement in the Identification of a Crack Tip Echo in Ultrasonic Inspection using Large Displacement Ultrasound Transmission

T Mihara[1] and H Ishida[2]

[1]Professor, Graduate School of Engineering, University of Toyama, JP
[2] Senior Researcher, Institute of Nuclear Safety System, Incorporated, JP

E-mail: mihara@eng.u-toyama.ac.jp

Abstract. As an improvement in the imaging technique for the low S/N ratio of crack tip echo in ultrasonic inspection, we developed the Subharmonic Phased Array Crack Evaluation (SPACE) system. Although the pulser of conventional SPACE can generate large, over tens of nm, displacement ultrasound at a crack, it is effective only for closed cracks and not effective for most industrial cracks. For general use of the SPACE system in industrial inspection, we need to develop larger displacement ultrasound incidence equipment considering crack openings of the order of several nm to sub μm. In this study, we developed a high voltage excitation SPACE and larger amplitude ultrasound incident to a crack using a high voltage proof transducer. The suitability of the developed system was investigated using typical models of cracks.

1. Introduction

Accurate sizing in industrial inspection is a particular requirement of ultrasonic inspection to guarantee the strength of aging infrastructures. All the industrial ultrasonic inspection techniques, such as the tip echo method, time of flight diffraction (TOFD) and phased array, have used the crack tip diffraction echo. Thus, the measurement accuracy of all these industrial methods depends on the detection of the industrial crack tip. Although almost all crack tip detection using conventional ultrasonic inspections is accurate, it is difficult to detect the crack tips of some cracks because of the low signal to noise ratio (S/N ratio) of the crack tip diffraction echo caused by crack closure or by the large ultrasonic scattering noise, and so on. In these cases, correct judgment of the crack tip diffraction echo is sometimes difficult even for skilled engineers. Thus, new techniques are demanded to assist in the reliable determination of a crack tip. For this purpose, we developed the Subharmonic Phased Array for Crack Evaluation (SPACE) system [1-6]. Imaging in the SPACE system uses synthetic aperture procedures and additional digital filtering by improving a commercial phased array system. As a result, the fundamental image (for an incident frequency of ω, the ω image) and the subharmonic image (the $\omega/2$ image) are able to be obtained independently using the SPACE system. We have demonstrated the use of the SPACE system in sizing the crack for a fatigue crack and a Stress Corrosion Crack (SCC) especially for a closed crack, which until now has been difficult to detect using conventional ultrasonic inspection techniques.

In this study, we investigated the validity and limitation of the existing SPACE system for the measurement of simulated industrial cracks, which were provided as model specimens for a nuclear power generator. Furthermore, after the development of the high voltage SPACE system, as well as improving the transmit transducer, we investigated the validity of using the new SPACE system to extend the applicable limitations for generating subharmonic ultrasound at industrial cracks in the existing SPACE system.

2. SPACE Measurement System

The SPACE system is the first imaging system to use subharmonic ultrasound, which is able to obtain the fundamental and subharmonic images independently. Its schematic diagram is shown in Fig. 1. Subharmonic ultrasound is considered to be generated by clapping only at crack surfaces, which is the advantage in identification procedures of the crack tip, and is a more reliable indicator in accurate inspection than superharmonic ultrasound because large superharmonic ultrasound generates images even at ultrasonic wedges and couplants other than at a crack. It combines aperture synthesis procedures and digital filter processing using the commercial phased array system of PAL-3. The ultrasonic transmitter of the SPACE system used a high voltage pulser with a maximum excitation of 2200 V, which is standard equipment in nonlinear ultrasound measurement [3].

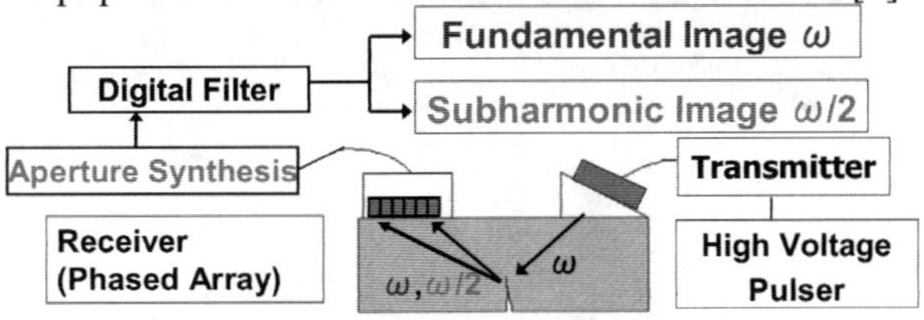

Fig.1 Schematic diagram of SPACE

To use reliable fundamental and subharmonic images in ultrasound measurement, narrow band burst excitation is required. However, when we use commercial or hand-made transducers for subharmonic ultrasound using high voltage burst pulses, these transducers can be damaged by heating and electrical shorting because of the high level of acoustic energy.

After the trials resulting in damaged transducers, we decided to use the transmit pulser below 500 V excitation voltage as the standard measurement condition for quantitative measurement conditions in our laboratory. An example of measurements, in these conditions, on aluminium alloy with a closed fatigue crack, is shown in Fig. 2 [3]. Because of crack closure, a crack tip echo was not detected by linear ultrasound of the fundamental image and, in contrast, a subharmonic image did detect the crack tip.

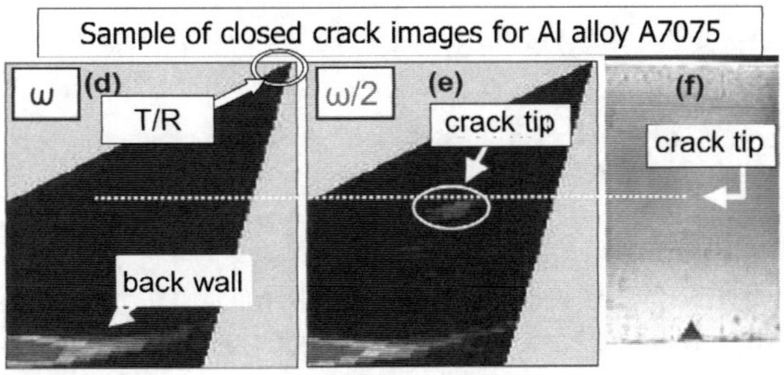

Fig.2 Example of the advantage of SPACE on aluminium alloy

As shown here, the SPACE system has been effective in the detection and accurate sizing of an industrial closed crack caused by residual compression stress. Conversely, many industrial cracks can be detected fairly accurately using conventional linear ultrasonic inspection methods, though some of their S/N ratios tend to be lower. If almost all industrial cracks are considered to be open, using the existing SPACE measurement system may be limited to only fully-closed cracks [1-6]. These limitations on the use of the existing SPACE techniques will become an impediment to extending the use of SPACE in industrial inspection fields.

3. Evaluation of industrial cracks using the existing SPACE system

To investigate the validity of the SPACE system, several specimens with various fatigue cracks and SCCs were measured using the SPACE system. Series A, the austenitic stainless steel SUS316 specimens (weld material is an Inconel alloy) with multiple SCCs, and series B, the austenitic stainless steel SUS316 weld specimens with a SCC, were prepared. In the future, we plan to slice cross-sections of the specimens to accurately measure crack size, however in this study, the depth of the SCCs was an estimated value. An excitation voltage of a burst of three waves at 400 V was applied as a standard condition for qualitative analysis using a single PZT transducer. An example of the measurement on specimen A is shown in Fig. 3. The figure on the left shows the fundamental image and on the right the subharmonic image. For SCC-1 the fundamental image (a) detected the crack tip, and the subharmonic image (b) did not. From this behaviour SCC-1 is assumed to be an open crack. However, for SCC-2, the depth of the crack tip detected by the subharmonic image was greater than the estimated crack depth and the fundamental image did not detect the crack tip. From this behaviour SCC-2 is assumed to be a closed crack.

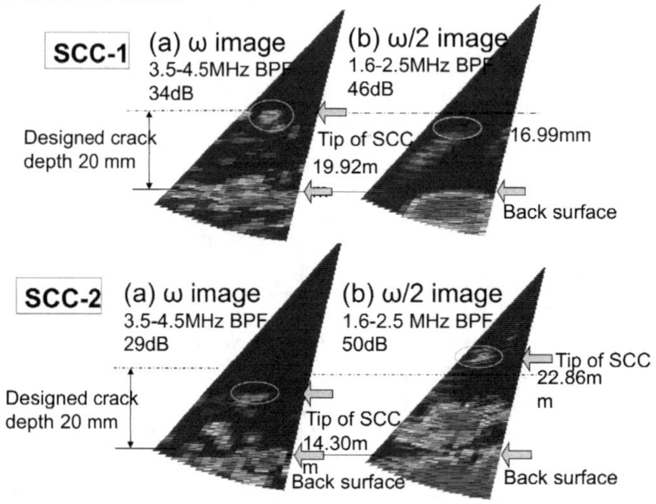

Fig.3 Examples of the conventional SPACE system (excitation voltage of a burst of three waves at 400 V)

Regardless of the SPACE system measurements of many other cracks, the existing SPACE system and subharmonic ultrasound is effective only for a closed crack and not for an open or semi-open crack.

4. Large amplitude ultrasound transmission techniques

Although the quantitative mechanism for generating subharmonic ultrasound at a crack has not been investigated until now, the incident ultrasound displacement must be greater than the opening displacement of a crack to generate the subharmonic ultrasound qualitatively [7, 8]. Ultrasound almost ten times greater in amplitude than in conventional ultrasonic inspection has been required for the generation of non-linear ultrasound. Thus, an existing SPACE system chose a high voltage burst pulser with a maximum excitation of 2200 V; however damage to the electrodes and electric-wire bonding of the transducer will occur at repeated voltage burst excitations beyond 500 V. In this research, the structures in the transducer were improved to withstand large voltage burst wave excitations of up to 1100 V for stable subharmonic ultrasound measurement. The details of the structure of the improved transducer are shown in Fig. 4. The various causes of damage to the transducer from high voltage excitation were 1) heating of an electric impedance matching coil, 2) electrical shorts between electrodes and 3) breaks in the connecting structure between electrodes and the electric wires. To avoid damage, the structure of the electrode for a single PZT element was improved by eliminating the coil of the matching circuit for the measurement using the SPACE system.

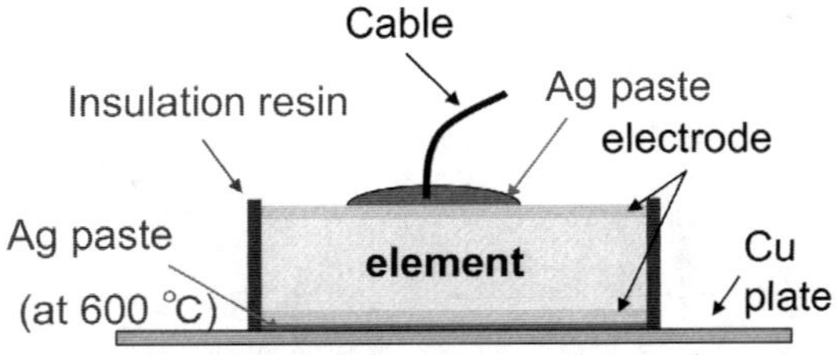

Fig.4 Structure of the improved transducer

Furthermore, instead of the silver/resin paste used in a commercial transducer, a new higher strength silver paste fabricated for operation at a high temperature was used. Because the fabricated temperature of the process is 600°C, which is higher than the Curie point of PZT, a repolarization process with a DC voltage of 1000 V/mm was applied for 30 min at 200°C after the fabrication of the transducer. Before the measurement using this transducer, the reliability of the improved structure for high voltage excitation was investigated using continuous bursts of 10 waves of excitation up to 1100 V for 50 h. The measurement results for the 2 MHz and 5 MHz transducers are shown in Fig. 5. Comparing the high voltage proof tests for transducers of the existing structures and of the improved structures, we confirmed the reliability of the improved transducer. The high voltage new SPACE system combining the improved 5 MHz transducer was applied to several cracks.

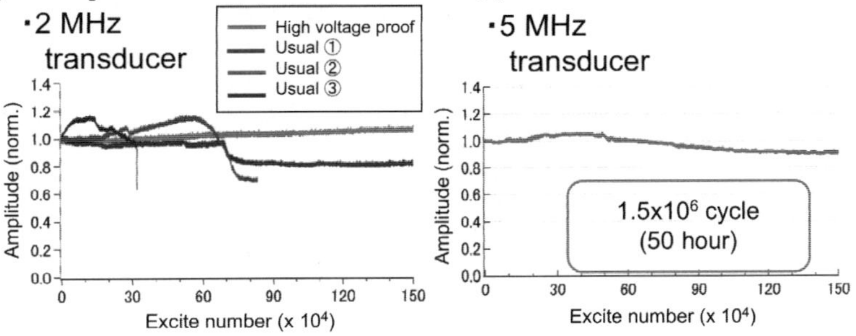

Fig. 5 High voltage proof tests for the improved transducers

5. Application of the high voltage SPACE system for open cracks

We increased the excitation voltage for an incident transducer of SPACE from the existing level of 400–1100 V for several cracks. An example of the results is shown in Fig. 6 for SCC-1 in specimen B. In Fig. 6, the tip echo of crack was clearly detected when using a 400 V excitation for the fundamental image, and the amplitude of the crack tip echo increased linearly with excitation voltages of up to 1000 V. However, the absolute amplitude of the tip echo for this crack was almost 10 dB lower than that for other open cracks for all the excitation voltages.

Conversely, in the corresponding subharmonic images, the crack tip echo was not detected using 400 and 500 V excitations similar to the conventional measurements in the cases of open cracks. However at an excitation voltage of 600 V a crack tip echo appeared in the subharmonic image and the amplitude of this echo increased with excitation voltages up to 1000 V. Because the location of the crack tip for the subharmonic image was the same as that for the fundamental image, the appearance of a subharmonic crack tip image could assist the reliable determination of a crack tip with the fundamental image measurement.

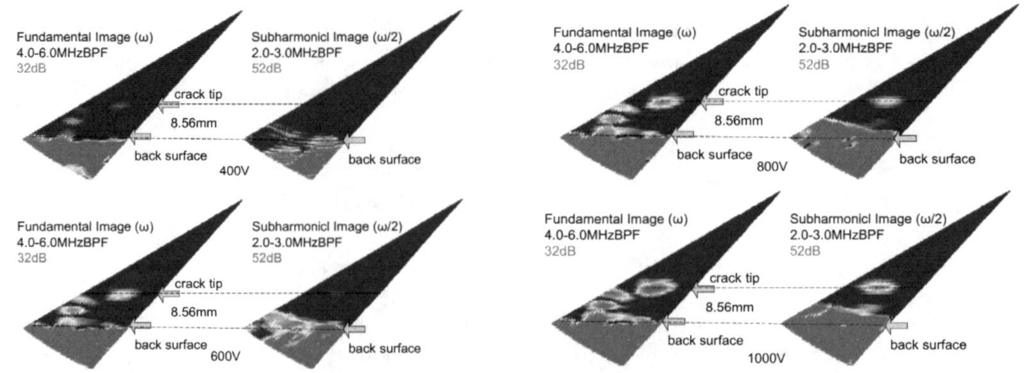

Fig.6 Crack tip images from the improved SPACE (Specimen B, SCC-1)

6. Considerations on the experiment and future work

To investigate whether the subharmonic image at the crack tip in Fig. 6 was not a leak of the fundamental ultrasound in the filtering process of SPACE, the initial waveform of the RF echo at 1000 V excitation before filtering was subjected to a wavelet transform as shown in Fig. 7. The frequency components of the crack tip echo had two independent peaks of fundamental and subharmonic frequencies. Thus, we confirmed that the subharmonic image at the crack tip in Fig. 6 was not a leak of the fundamental ultrasound but an apparent subharmonic ultrasound.

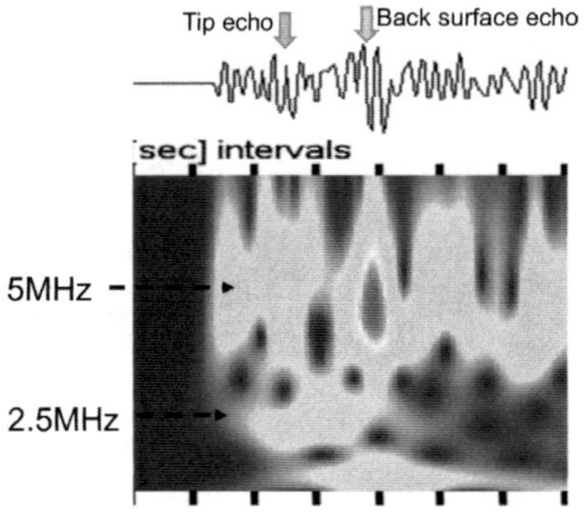

Fig. 7 Wavelet transform for the RF waveform of 1000 V excitation for SCC-1 in specimen B.

The amplitude of the crack tip echoes in the fundamental and subharmonic images for SCC-1 are summarized in Fig. 8. The noise level of the measurement was determined using the amplitude of the scattering noise from the microstructures of the welding specimen of austenitic stainless steel.

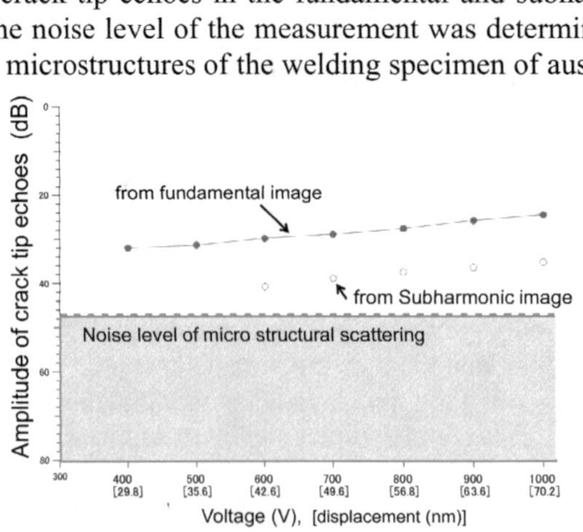

Fig. 8 Amplitude of the crack tip echoes in the fundamental and subharmonic images in Fig. 6.

At an excitation of 400 V, as our standard condition for the existing SPACE system measurement, SCC-1 was considered to be open because the crack tip echo was detected only in the fundamental image. Moreover, because the amplitude of the fundamental crack tip echo for SCC-1 was 10 dB lower than any other fully-opened crack, we estimated, on inspection, that SCC-1 was a semi-opened crack with a lower S/N ratio. The crack tip echoes increased linearly with the increase in excitation voltage amplitude. At an excitation of 600 V, the crack tip echo appeared in the subharmonic image for SCC-1 and its amplitude also linearly increased with the increase in excitation voltage. Thus, we conclude the following: (1) the other fully-opened cracks could be detected clearly using fundamental linear measurements with adequate S/N ratios, and (2) for the semi-opened crack of SCC-1, the detectability and reliability of crack tip detection decreased because of the lower S/N ratio of the crack tip echo. For this semi-opened crack, subharmonic ultrasound was generated when the large amplitude ultrasound with higher than 600 V excitation was incident to the crack for SPACE system measurement. Ultrasonic detectability of crack tips for SCC-1, with a lower S/N ratio in crack tip detection than for other fully-opened cracks in conventional inspection, was improved using the subharmonic image. Furthermore, for the closed crack, subharmonic measurement could be effective as shown in SCC-2 of Fig. 3(b).

In the future, when larger amplitude ultrasound is incident to the crack, we may generate the subharmonic waves even for more opened cracks. In addition, for closed crack measurement, larger amplitude ultrasound will be effective in improving the tip detection of tightly closed cracks caused by the larger residual compression stress from welding processes. Thus, although the maximum displacement from the transmit transducer might be a limitation, we believe that larger amplitude-incident ultrasound will be one of the key techniques in the SPACE measurement system for use in industrial fields

References

[1] Mihara T, Nomura S. Akino M and S and Yamanaka K 2004 Relationship between crack opening behavior and crack tip scattering and diffraction of longitudinal waves *Materials Evaluation* **62** 943-47

[2] Yamanaka K, Mihara T and Tsuji T 2004 Evaluation of closed cracks by analysis of subharmonic ultrasound *Insight* **46** (11) 666-70

[3] Ohara Y, Mihara T, Sasaki R, Ogata T, Yamamoto S, Kishimoto Y and Yamanaka K 2007 Imaging of closed cracks using nonlinear response of elastic eaves at subharmonic frequency *Appl. Phys. Lett.* **90** 011902

[4] Ohara Y, Yamamoto S, Mihara T and Yamanaka K 2008 Ultrasonic evaluation of closed cracks using subharmonic phased array *JJAP* **47** (5) 3908-15

[5] Ohara Y, Yamamoto S, Endo H, Mihara T and Yamanaka K 2008 Accurate measurement of closed cracks using subharmonic phased array *Nonlinear Acoustics- Fundamentals and Applications* (18th ISNA) **18** 557-60

[6] Ohara Y, Endo H, Mihara T and Yamanaka K 2009 Ultrasonic measurement of closed stress corrosion crack depth using subharmonic phased array *JJAP* **48** (7) 07GD01-1-6

[7] Mihara T, Japanese Patent 2012 No. 2012-88113

[8] Mihara T, Kamimura T, Miura Y and Tashiro H, "Development of high amplitude ultrasonic laminated transducer system for ultrasonic inspection 2012 Proceeding of *IWPMA 2012* 61.

Acknowledgments

This work was partially supported by a Grant-in-Aid for the Innovative Nuclear Research and Development Program (No.120804) from the Ministry of Education, Culture, Sports, Science and Technology of Japan.

High Sensitivity EMAT System using Chirp Pulse Compression and Its Application to Crater End Detection in Continuous Casting

Y Iizuka[1] and Y Awajiya[2]

Steel Research Laboratory, JFE Steel Corporation
[1]1-1, Minamiwatarida-cho, Kawasaki-ku, Kawasaki, 210-0855, Japan
[2]1, Kawasaki-cho, Chuo-ku, Chiba, 260-0835, Japan

Email: y-iizuka@jfe-steel.co.jp

Abstract. A high sensitivity EMAT system using chirp pulse compression technique was developed. The system uses a high power gated amplifier having 2kVpp output to transmit chirp waves. Pulse compression of the received signals are performed digitally in a PC after amplification and analog-to-digital conversion. A 20dB improvement of the signal-to-noise ratio was achieved by chirp pulse compression and synchronous averaging. A new surface cooling technique was also developed to improve the signal amplitude of the bulk shear wave with hot steel, and its effectiveness was demonstrated. An actual plant test of crater end detection by the developed EMAT system was conducted at a continuous caster, and clear detection by non-contact EMATs was achieved.

1. Introduction

In the steel industry, non-contact ultrasonic testing methods have been expected as in-line nondestructive evaluation techniques for hot processes. Crater end detection in continuous casting is one example of this need. It is required in order to improve casting productivity and slab internal quality. Electromagnetic acoustic transducer (EMAT) techniques had been tried previously [1,2,3], but due to the low sensitivity of conventional EMAT techniques, liftoff had been limited to less than 2mm. Since a touch roll is used to maintain this narrow liftoff to a hot material whose surface temperature is about from 800°C to 1000°C in in-line operation, this technique approaches contact testing, resulting in low durability of the EMATs.

This paper proposes a new high sensitivity EMAT technique. In order to improve the signal-to-noise ratio (SNR), the authors developed a chirp pulse compression EMAT system. This system transmits chirp signals by using an arbitrary waveform generator and a high power gated amplifier having 2000Vpp output. Pulse compression of the received signals is performed digitally by software in a PC. The SNR was improved by more than 20dB by chirp pulse compression in combination with 16 synchronous averages. As a result, liftoff was improved by +5mm, achieving perfect non-contact measurement. In addition, signal enhancement by the magnetostriction effect by surface cooling was verified when using a bulk shear wave with hot material.

An actual plant test of crater end detection by the developed system was conducted at a continuous caster, and clear detection by perfect non-contact EMATs was successfully achieved.

2. High sensitivity EMAT system using chirp pulse compression

Signal processing is an effective way to improve the SNR of ultrasonic testing signals. In particular, synchronous averaging is often used. However, the number of averages cannot be increased greatly in in-line measurement because the material is moving and real-time measurement is required. Therefore, the pulse compression technique was applied to obtain a high SNR and short pulse within one pulse. Chirp wave is used as the transmitting pulse. The chirp wave has the features of a low side-lobe and easy setting of the frequency band compared to the coded pulse technique. Chirp pulse compression has been used previously for flaw detection of pipes [4]. The authors developed a high power system to enable use of EMAT.

Figure 1 shows the high power EMAT system. A high power gated amplifier is used to transmit high power chirp signals. This device is a custom-ordered GA-5000 manufactured by Ritec Inc. The output voltage is about 2kVpp at 1% duty. The received signals are amplified by a low noise operational amplifier and digitized by a digital oscilloscope, after which pulse compression is performed by software. Synchronous averaging is performed in combination with this process by the digital oscilloscope. The number of averages is set at 16, which is allowable in in-line measurement.

The procedure of the signal processing is shown in Figure 2 as an example of a simulation. The chirp wave generated by Eq. 1 is used as the transmitting pulse. After synchronous averaging by Eq. 2, correlation is carried out for the pulse compression by Eq. 3. As the auto-correlation is like a delta function, the pulse width of the chirp signal is compressed by the cross correlation between the received signal and the chirp signal. Random noises are suppressed because the cross correlation between random noise and the chirp signal is low. Thus, a short and high-SNR signal is obtained.

Figure 3 shows an experimental result of SNR improvement at room temperature. In this test, a 250mm thick stainless steel block is used as simulated hot carbon steel. A transmission method with bulk shear wave EMATs is used. The chirp pulse was compared with a tone burst pulse as a conventional technique. The conditions of the transmitting signals are shown Table 1. SNR is defined as the signal amplitude to the noise amplitude ratio. The amplitude is measured as the maximum value in the gate as shown in the Figure 3(c).

A result is shown in Figure 3. The signal amplitude of the conventional burst wave decreases with increasing liftoff, and it becomes difficult to distinguish the signal from noise at 7mm, even though the SNR was 20dB at 2mm liftoff. On the other hand, with the chirp pulse compression, a 15dB SNR is obtained at 7mm liftoff. A 24dB SNR is obtained with 16 synchronous averages. As the result, a 5mm liftoff improvement is achieved.

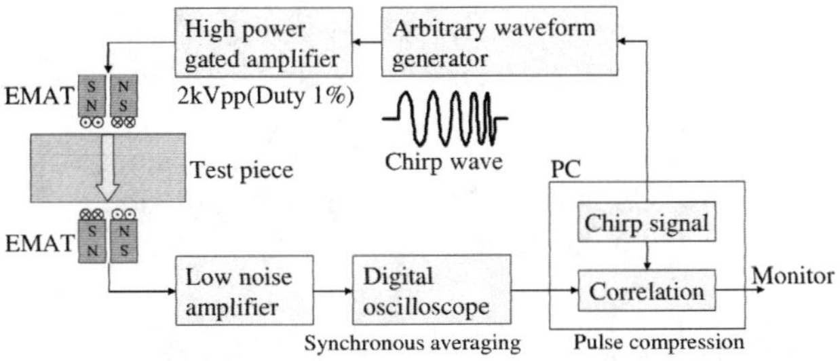

Fig. 1 Configuration of high sensitivity EMAT system

Table 1 Transmitting signals

Tone burst	
Center frequency	300kHz
Cycles	2
Chirp wave	
Center frequency	300kHz
Band width	500kHz
Pulse width	40 μ s

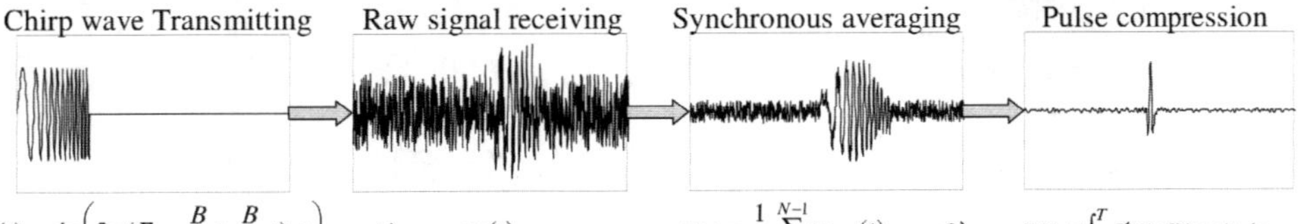

$$st(t) = \sin\left(2\pi(Fc - \frac{B}{2} + \frac{B}{2T}t)\cdot t\right) \quad \cdots 1)$$

$sr_i(t)$

$$x_i(t) = \frac{1}{N}\sum_{k=0}^{N-1} sr_{i-k}(t) \quad \cdots 2)$$

$$y_i(t) = \int_0^T st(\tau)\cdot x_i(t+\tau)\cdot d\tau \quad \cdots 3)$$

Fc : Center frequency
B : Band width
T : Pulse width

N : Number of average
i,k : Pulse number

Fig. 2 Signal processing procedure

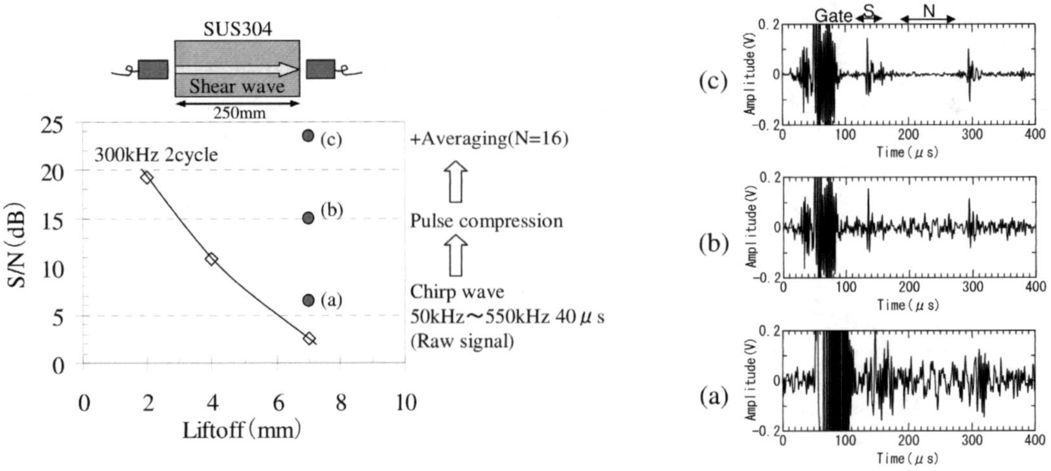

Fig. 3 Experimental result of S/N improvement

3. S/N improvement by surface cooling for hot steel

One feature of the EMAT technique is easy generation of a bulk shear wave. Shear wave is considered to be an appropriate mode for crater end detection as this type of wave is not transmitted to liquids (in this case, molten steel). However, it has been reported that the signal of the shear wave becomes low at more than 750°C in the case of carbon steel [5]. Therefore, the authors attempted application of a surface cooling technique to enhance the signal of the bulk shear wave EMAT based on the facts that the critical temperature is around the transformation temperature, and magnetostriction contributes to the sensitivity of the shear wave EMAT compared to the Lorentz force in the case of carbon steel [6]. Surface cooling reduces the surface temperature, enabling magnetism, and as a result, the bulk shear wave can be generated and detected by the magnetostriction effect rather than by the Lorentz force, which is used conventionally with hot steel.

In this test, a 100mm thick carbon steel block, which had been heated to 1000°C, and bulk shear wave EMATs were used. The EMATs were cooled with water to protect against heat damage. A tone burst wave was used as the transmitting pulse. The center frequency was 300 kHz and the number of cycles was 3. The temperatures were measured with thermocouples at the center and 3mm depth of the sample. The amplitude of the transmitting signal was measured using the gate S as shown in the Figure 3(c).

Figure 4 shows an experimental result of the surface cooling test. Although the amplitude of the transmitting signal was low at first, from 25 seconds after cooling, the signal was enhanced about 6dB. Thus, signal enhancement by the magnetostriction effect was confirmed.

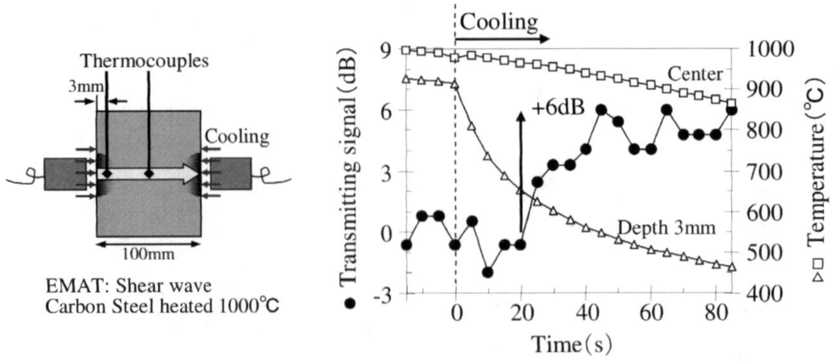

Fig. 4 Experimental result of signal enhancement of shear wave by surface cooling for hot steel

4. Actual plant test of crater end detection

An actual plant test of crater end detection was conducted at a bloom continuous caster by using the developed EMAT technique. The shear wave transmission method [2,3] was used. In case of complete solidification, the transmitting shear wave signal is detected. However, when the crater end reaches the position of the EMATs, the transmitting shear wave signal disappears, as the signal is not transmitted to the liquid steel. A pair of EMATs was installed near the end pinch rolls with 5mm liftoff from the sensor box. The sensors were thermally protected by a 1mm thick waterway in the sensor box. A surface cooling device to enhance the signal of the bulk shear wave was installed just upstream from the EMATs. The surface temperature was about 900°C before cooling and less than 700°C after cooling. The thickness of the bloom was about 250mm.

Figure 5 shows the trend of the transmitting shear wave signal during a casting speed change test. The amplitude is measured in the gate shown on the figure. After the casting speed was increased, the transmitting shear wave signal decreased. Crater end detection was clearly observed at point A in the figure, as the signal completely disappeared. In addition, after the casting speed was decreased, the transmitting shear wave signal appeared at point B, indicating movement of the crater end to the upstream side. The SNR of the transmitting shear wave signal was significantly higher, displaying an improvement of more than 20dB. As this experiment demonstrated, the crater end can be clearly detected with perfect non-contact sensors by using the developed EMAT system.

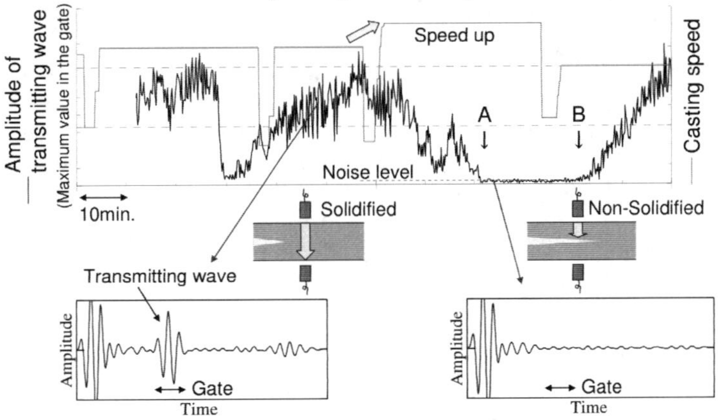

Fig. 5 Experimental result of crater end detection in actual continuous casting

5. Conclusions

A high sensitivity EMAT system using chirp pulse compression was developed. The results are summarized as follows.

1) By using chirp pulse compression and synchronous averaging, a short, high SNR pulse can be obtained at more than 5mm liftoff.

2) A surface cooling technique is effective to enhance the signal of the bulk shear wave EMAT when the system is applied to hot carbon steel, even when the internal temperature is more than 900°C.

3) Clear detection of the crater end by perfect non-contact EMATs was successfully achieved in an actual plant test at a continuous caster.

References

[1] K. Kawashima et al., Testu to Hagane 67(9), 1515-1522 (1981)
[2] H. Takada et al., Current advances in materials and processes-ISIJ 2, 1214 (1989)
[3] P.W.Manos et al., 9th Process Technology Conference proc., 139-150 (1990)
[4] Y. Iizuka, Insight 40(4), 282-285 (1998)
[5] K. Kawashima, Hihakaikensa 34(11), 796-803 (1985) in Japanese
[6] H. Ogi, J Appl Phys 82(8),3940 (1997)

Journal of Physics: Conference Series **520** (2014) 012012 doi:10.1088/1742-6596/520/1/012012

Evaluation of the Surface Roughness using AE method with Air Blowing

T Yasuda[1], S Takata[1], T Hino[1] and K Yoshida[2]

[1] Department of Mechanical Engineering, Anan National College of Technology, Anan-city Tokushima 774-0017, Japan

[2] Department of Mechanical Engineering, Faculty of Science and Engineering, Tokushima Bunri University, Sanuki-city Kagawa 769-2193, Japan

E-mail: yasuda@anan-nct.ac.jp

Abstract. This study aims to find the development for the evaluation of the surface roughness by the Acoustic Emission (AE) method with air blowing. We paid attention to the AE wave due to air blowing on the specimen plate with different surface roughness. The relationship between the AE wave and surface roughness of specimen plates was investigated. As the result, there is large and continuous difference in the Root Mean Square (*RMS*) value of their AE waveform. The *RMS* value decreases by increasing of the surface roughness of specimen plates. It suggested that this characteristic has the possibility to establish a new method of non-destructive surface roughness testing.

1. Introduction

Acoustic emission (AE) is defined as the ultrasonic wave phenomenon due to rapid release of strain energy during the fracture, crack generation or growth, deformation and so on in some kinds of structures or solid materials. By the sensing of AE wave and collecting of AE parameters, we can analyze the specific phenomenon that worked as the AE source. The AE analysis has the possibility to determine the behavior of materials such as martensitic transformation, micro cracking or some kinds of phenomenon. Furthermore, the AE occurs during gas leakage from artificial defects, the pinholes with several sectional planes, on the steel pipe [1, 2]. It means the AE was also generated due to air blowing. In this research study we paid attention to the latter phenomenon, the AE due to air blowing, for developing a new method of the surface roughness testing.

Krogstad P. A. and Antonia R. A. were summarizing some researches and introducing them about the normal stresses in boundary layer due to flows along the wall; the normal stress is decreased by increase of the ordinate in boundary layer [3]. Considering their research, we thought that the AE source during air blowing is the interaction between normal stresses and wall in boundary layer of the flow.

In this research we investigated a tendency of the relationship between the AE waves due to air blowing on the specimen plates and theirs surface roughness. Four aluminum specimen plates with different surface roughness were prepared and constant airflow was blown to them in the experiment. During the experiment, the AE waves can be observed from the specimens. We compere the Root Mean Square (*RMS*) value of the AE waves due to four types of specimen. From the results it was suggested that this relationship investigated has the possibility of a new development for the evaluation of the surface roughness by the AE method with air blowing.

Journal of Physics: Conference Series **520** (2014) 012012 doi:10.1088/1742-6596/520/1/012012

2. Methodology

2.1. Experimental setup

Figure 1 shows the schematic diagram of experimental setup. We blew the compressed air (from Hitachi Bebicon, 0.7 MPa) to the specimen plate. At the same time, the AE monitoring was conducted. The front surface of the specimen plate was set at an angle of 45 degree to the direction of air nozzle. The AE sensor (Fuji ceramics M5W) was attached on the back surface of the specimen plate. The detected AE waves were amplified by the pre-amplifier (NF9913, 40 dB) and sent to the digital oscilloscope (GW Instek GDS-1072A-U). After the experiment, detected AE waves were investigated on the personal computer.

In fact about the experimental setup, the air nozzle and the specimen plate were fixed strongly by fixing tools. During the experiment, blowing air to the specimen plate, the flapping motion of specimen plate was not observed. Furthermore, we utilized the pencil lead break testing for proofing the noise detection. The pencil lead break testing is one of the methods to simulate the burst AE wave from the fracture of the brittle graphite lead in a suitable setting. It is often used as the method for excitation of the artificial AE source [4]. In this study, the setting and the graphite lead in Figure 2 was prepared. Breaking a diameter = 0.3 mm and length = 3 mm pencil lead on the surface at several spots of the fixing tools. We checked no AE wave detection from the fixing tools. When the pencil lead was broken only on the specimen plate, the AE wave was detected.

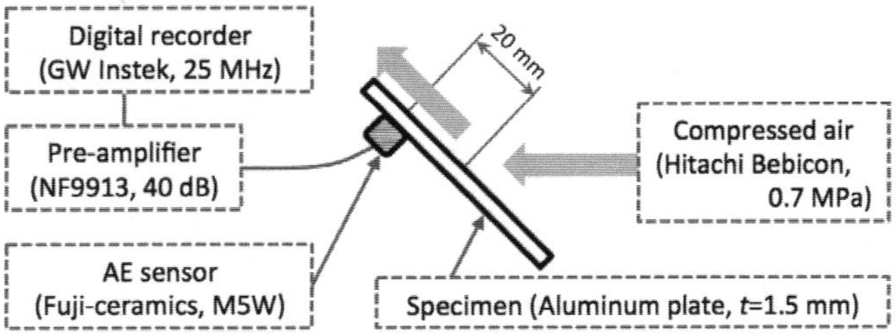

Figure 1. Schematic diagram of experimental setup

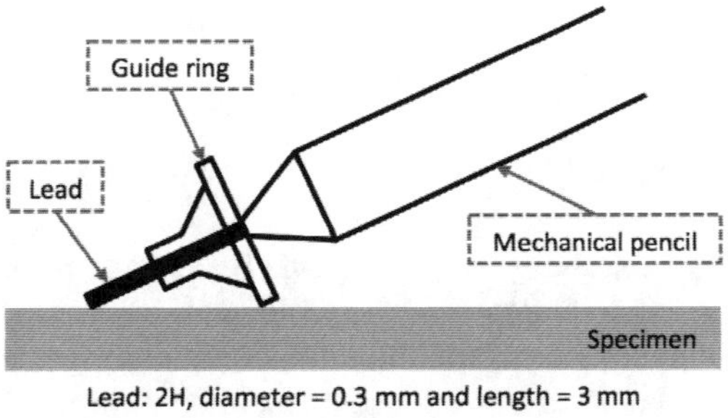

Figure 2. Pencil lead break testing

2.2. Specimen

Four types of specimen plates (Aluminum plate, thickness = 1.5 mm) were prepared with different surface roughness. The first specimen plate has untreated surface. The second one is sanded using #800 sanding paper. The third and fourth one are filed using smooth- and rough-cut files. Each back surface of the specimen plates have untreated surface.

The surface roughnesses of the four specimen plates were measured utilizing the stylus type surface roughness tester (SURFCOM 130A). Each calculated average roughnesses were Ra = 0.223, 1.50, 1.91 and 3.04 μm. The same experiment of air blowing and AE monitoring above-mentioned were conducted to these four specimens.

3. Result and Discussion

Figure 3 shows the typical AE waveforms. During the experiment, the continuous AE wave was observed on the back surface of the specimen plates. The third line of Table 1 shows the Root Mean Square (*RMS*) value of their AE waveforms. The *RMS* value means the average power of the AE waveform. It is given by next formula:

$$RMS = \sqrt{\frac{1}{\tau}\int_{t}^{t+\tau} V^2(t)dt}$$

where $V(t)$ is the voltage from pre-amplifier and τ is the time range of the AE waveform.

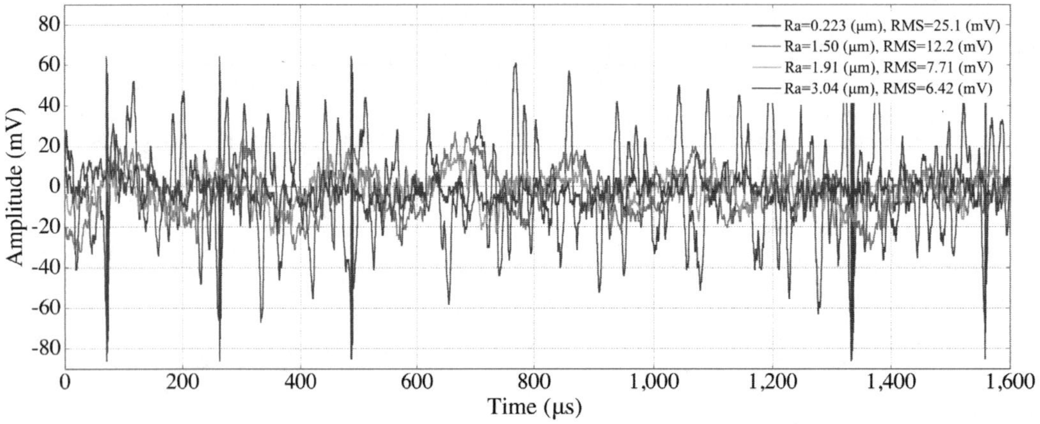

Figure 3. Typical AE waveforms during air blowing experiment

Table 1. Surface of the specimen

Specimen number	No. 1	No. 2	No. 3	No. 4
Ra of specimen (μm)	0.223	1.50	1.91	3.04
RMS of experiment (mV)	25.1	12.1	7.71	6.42
RMS of Pencil lead break (mV)	18.3	17.0	19.2	20.3

In Table 1, the *RMS* value decreases by increasing of the surface roughness of specimen plates. And there is the behavior of continuously decreasing in *RMS* value between smooth and rough surface specimen plates. On the other hand, the fourth line of Table 1 shows the *RMS* value of the AE waveforms due to the pencil lead break testing on each specimen plates. The pencil lead break testing was conducted using same suitable setting above-mentioned section 2.1. The AE waves and the *RMS* value of the pencil lead break testing were almost same. Therefore from these results, it is thought that the difference of *RMS* values due to air blowing has remarkable behavior in comparison with the result of the pencil lead break testing. We supposed that this characteristic possess the ability to grow as the new method of non-destructive surface roughness testing.

Figure 4 is the AE waveforms in frequency domain of four AE waveforms shown in Figure 3. It seems that only untreated surface specimen, the smoothest surface in four specimen plates, has stronger high frequency characteristic especially range of over 400 kHz. It suggested to some different phenomenon is included as the AE source. We expect about the effect of the flow separation. The rougher surface makes easy to separate the airflow from the object. Therefore usually for preventing the flow separation, the shape of objects is streamlined smoothly [5]. Suppose that the normal stress in boundary layer of the flow was the AE source during air blowing in this research, it thought that the AE source was lost when the blowing airflow was separated from the rougher specimen plate and the ordinate of boundary layer was increased. Thus it thought that the increases of surface roughness of specimen plates decreased the *RMS* value of AE wave.

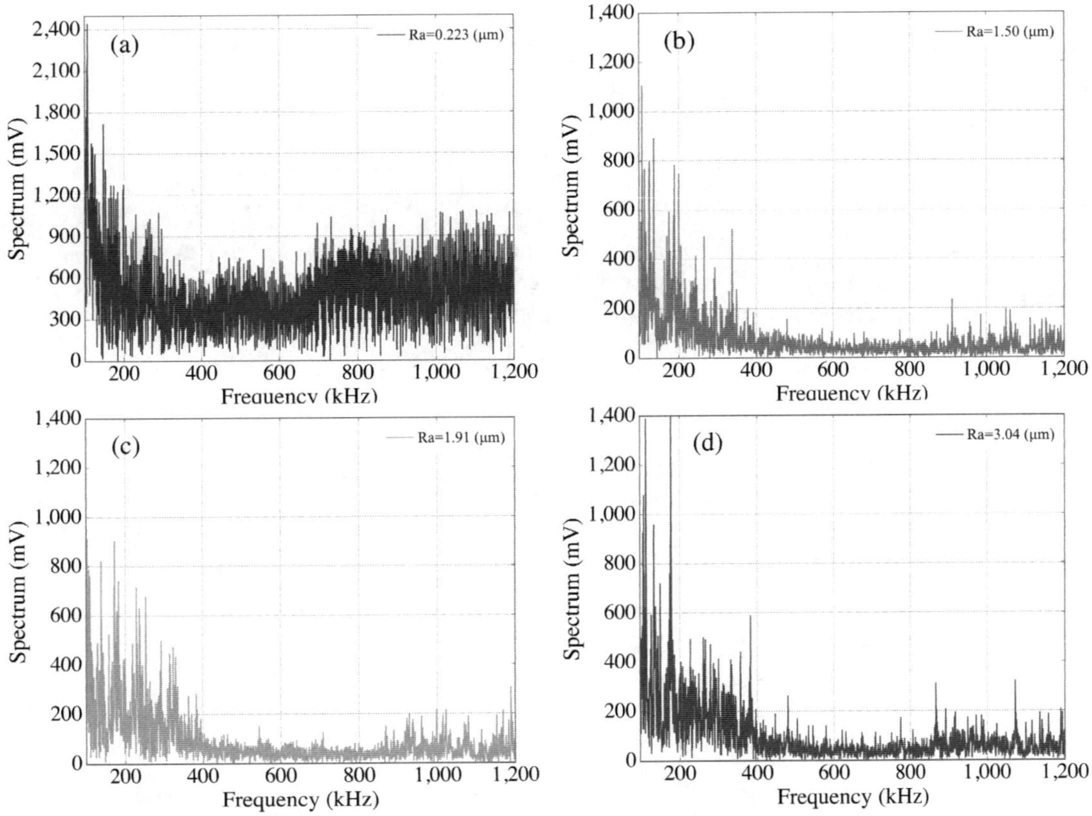

Figure 4. AE waveforms in frequency domain during air blowing experiment (a) *Ra* = 0.223 μm, (b) *Ra* = 1.50 μm, (c) *Ra* = 1.91 μm and (d) *Ra* = 3.04 μm

4. Conclusion

In this study, we conducted the AE monitoring during the air blowing to the four kinds of specimen plate with different surface roughness and investigated the *RMS* value of AE waveforms observed. There are large changes in *RMS* value between smooth and rough surface specimen plates. The *RMS* value decreases by increasing of the surface roughness. Furthermore in the case of the pencil lead break testing on each specimen plate, the difference of the *RMS* values of artificial AE wave was small.

References

[1] Laodeno R N, Nishino H, Yoshida K 2008 *Materials Transactions* **49**-10 2341-2346
[2] Laodeno R N, Yoshida K 2013 Non-destructive Testing (NDT) Technique, Acoustic Emission Analysis of Gas Leak on Pipe *Lambert Academic Publishing*
[3] Krogstad P A, Antonia R A 1999 *Experiments in Fluids* **27** 450-460
[4] e.g. Shiotani T, Takada Y, Watanabe T and Ohtsu H 2012 *Progress in Acoustic Emission* **XVI** 1-6
[5] Nakamura K et al. 1997 Hydrodynamics (in Japanese) *Power Company*

Temperature Dependent Elasticity of a Ni_2MnGa Film as Measured by Impulsive Stimulated Scattering

J Sermeus[1], B Verstraeten[1], P Stoklasova[2], M Landa[2] and C Glorieux[1]

[1] Laboratory for Acoustics and Thermal physics, Department of Physics and Astronomy, KU Leuven, Celestijnenlaan 200D, 3001 Heverlee, Belgium
[2] Institute of Thermomechanics, Academy of Sciences of the Czech Republic, Dolejskova 5, 182 00 Prague 8, Czech Republic

E-mail: jan.sermeus@fys.kuleuven.be

Abstract. We report the results of an impulsive stimulated scattering measurement conducted on a 550 nm film of Ni_2MnGa on a MgO substrate, aiming at the direct experimental observation of the highly temperature dependent elasticity. The phase velocity of the surface acoustic waves excited on the surface of the coating-substrate shows an anomaly around the phase transition temperature of Ni_2MnGa, showing that it is possible to use impulsive stimulated scattering as a characterization tool to determine temperature dependent elasticity of a thin film.

1. Introduction

Impulsive stimulated scattering (ISS) is a laser ultrasonic technique used to excite and detect monochromatic surface acoustic waves (SAW) on a material. It has proven to be an useful tool in the characterization of bulk solids [1,2], free liquid surfaces [3], thin films [4] and multilayered structures [5]. A more extensive review of the technique, applications and the relationship of ISS to other laser ultrasonic methods, e.g. picosecond ultrasonics, is described in reference [6]. While it is possible to use the ISS technique in a transmission mode for the thermoelastic characterization of liquids [7], a reflection scheme was used here to characterize the opaque and highly reflective material. In this work, the phase velocity was measured for a SAW propagating along a crystal axis of a MgO single crystal substrate coated by a 550 nm Ni_2MnGa film.

Ni_2MnGa is a magnetic shape memory alloy (MSMA) with many interesting properties and with promising applications in MEMS devices. Since the phase velocity of SAW directly depends on the elastic properties of the material, anomalies in the SAW velocity reflect changes in the microstructure and are thus linked to the phase of the material. The temperature dependence of the SAW velocity was recorded in the vicinity of the magnetic and phase transition temperatures of the Ni_2MnGa film. Having experimental access to the temperature dependent elastic properties is therefore of great importance from a fundamental and application oriented point of view. A thorough temperature control of the investigated sample is required to correctly determine the transition temperatures. In this regard the question rises what effect the lasers have on the phase of the temperature sensitive coating. By comparing the observed and expected phase change temperature this question can be answered.

The conventional, thermally induced, shape memory effect (e.g. NiTi alloys) is based on a thermoelastic martensitic transformation (first-order solid-to-solid phase transition) from the cubic

Journal of Physics: Conference Series **520** (2014) 012013 doi:10.1088/1742-6596/520/1/012013

parent phase (austenite) to a less symmetric low temperature phase (martensite), or vice versa. The martensitic transformation is accompanied by a high macroscopic strain (up to 10%).

In Ni_2MnGa the magnetic shape memory effect is caused by a reorientation of the martensitic variants (twins) under an external magnetic field due to the high magnetocrystalline anisotropy of the martensite phase. Only modulated phases have a twinning stress low enough for the shape memory effect. The modulated (10M) martensite is often approximated by a tetragonal structure. However, recent studies [8,9] showed that the twinning structures are more complex, since there exist two types of twinning boundaries. Only one of which has a low twinning stress and is preferred for boundary motion. Detailed structural analysis and calculations in these references revealed the mechanisms for easy mobility of twinning interfaces explaining the low reorientation stress.

The sample under investigation was an epitaxially deposited Ni_2MnGa film [10] on a MgO substrate (100) by magnetron sputtering at a deposition temperature of 700K. The deposited film was in the paramagnetic austenite phase with epitaxial relation Ni_2MnGa (100)[110] || MgO (100)[100]. During cooling to room temperature, the martensite phases (mixture of 7M and NM) were reconfigured into hierarchical twin structures [10,11]. The martensitic transformation temperature for this sample was expected between 350-400K.

The Curie temperature of a bulk Ni_2MnGa crystal is a relatively stable value, T_c = 380-385K for different compositions [12]. However, for thin films T_c is lower and changes with deposition conditions (358K for this particular sample [13]). As the martensitic transition temperature T_{AM} is expected to be higher than T_c, it is not possible to detect the phase transformation temperature through magnetic measurements (e.g. SQUID, see [14,15]). An acoustical approach as presented in this work can provide a complementary tool in the characterization of the martensitic transformation in shape memory alloy thin films.

2. Experimental setup

2.1. Impulsive stimulated scattering setup

In the experimental setup of ISS, as depicted in figure 1, a 10 ps pulsed laser beam (1047 nm) is split into two beams by a first-order optimized diffraction grating. These two first-order laser beams are recombined at the sample surface to create an interference pattern with a specific fringe distance that depends on the diffraction grating and the optical imaging elements that are used. The wavelength of the excited wave was confirmed by registering and analyzing the Scholte wave, an interface wave between the solid sample and the air which travels at the speed of sound in air [16]. The pulsed pump laser has a peak power of 0.5 MW, which corresponds to an average power delivered at the sample of 5mW. At the sample surface a part of the incident light is absorbed, resulting in an impulsive heating. The sudden thermal expansion accompanied by this heating launches counter-propagating SAW with a wavelength corresponding to the interference fringe distance. This results in a standing wave pattern that oscillates at a central frequency f_{SAW}, corresponding to the ratio of the dispersive phase velocity c_{SAW} and the excited wavelength. In this work, the SAW propagated along the [100] direction of the (001) plane of the MgO substrate.

The detection part of the setup consists of a CW probe beam that exactly follows the optical path of the pump beam. At the sample surface the standing SAW pattern act as a temporally modulated diffraction grating for the two probe beams, which are thus partially reflected and diffracted. Recombining the two pairs of reflected and diffracted beams results in heterodyne detection [17] of the standing wave pattern. The intensity modulations were recorded by two Si pin-photodiodes (Hamamatsu S5973) and a high speed GHz amplifier (Femto series HSA). Subtracting the two heterodyne signals results in a doubling of the signal amplitude.

By recording the time signal on an oscilloscope, the central frequency f_{SAW} could be obtained after a fast Fourier transform analysis. In combination with the experimentally known wavelength of the SAW, the phase velocity could be determined for a set of temperatures ranging around the transition temperatures of the Ni_2MnGa coating.

SAW have the property that the penetration depth is comparable to the wavelength along the direction of propagation. In this work SAW with a wavelength of 10 μm were used to probe a 550 nm coating. It is clear from i.a. reference 5, where dispersion curves are shown for SAW propagating on coating-substrate systems, that even for large wavelength over coating thickness ratios the coating influences the SAW propagation.

2.2. *Temperature control*

The coating-substrate sample was mounted on a brass sample stage holding a calibrated Pt1000 temperature sensor and two 5W electrical resistances that served as external heaters. A PI-controller was used to automate and stabilize the temperature scan. In this way, the SAW phase velocity could be measured within a maximal uncertainty on the surrounding temperature of 0.1 K.

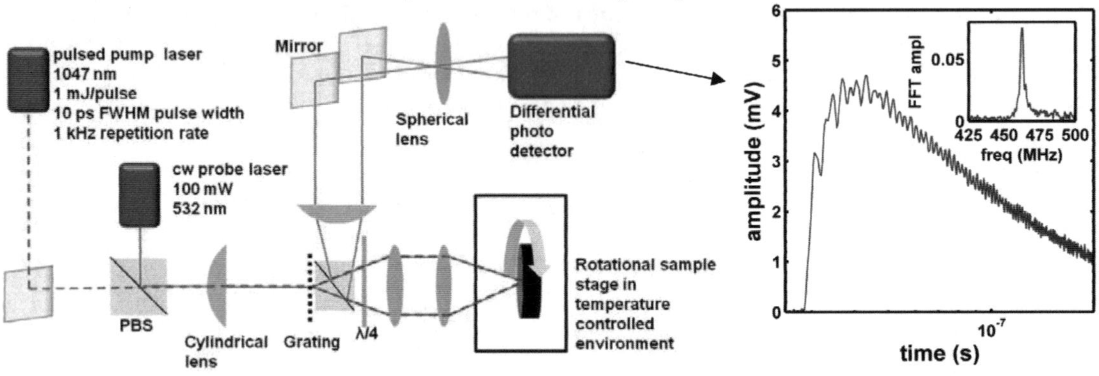

Figure 1. Schematics of the ISS setup with heterodyned detection. Pump and probe laser beams are superposed. The right figure shows a typical time signal with the FFT amplitude spectrum shown in the inset.

3. Results and discussion

In figure 2 the experimentally observed phase velocities of a SAW with a wavelength of 10 μm are shown as a function of the temperature. The transition from the ferro- to the paramagnetic state, which would show as a discontinuity in the first derivative of figure 2, was not observed as the temperature step was too rough. However, an anomaly in the phase SAW velocity around 370K is evident. This jump in the SAW velocity indicates a first order transition and it can be explained as the martensitic transformation. Moreover a narrow hysteresis between the forward (A/M) and reverse transition (M/A) is expected and observed.

The stiffening trend to the T_{AM} is an unexpected feature which might be explained by the influence of the substrate. A strain constrain, internal stresses or a rest of the austenite Ni_2MnGA might be present at the substrate-coating interface. Furhter investigation is needed.

4. Conclusion

In this article, we showed that the temperature dependence of the phase velocity of SAW propagating along the surface of a MgO substrate coated by a Ni_2MnGa film could be measured using ISS. An anomaly was observed around 370K and explained as the M/A phase transition temperature of the Ni_2MnGa coating. The effect of the lasers on the temperature and phase of the coating is therefore limited. Furthermore, this has work proven the feasibility of the exploited technique to experimentally obtain the temperature dependence of the elastic properties of magnetic shape memory alloys.

In future work the temperature dependence and magnetic field dependence of the thermal and elastic properties of the Ni_2MnGa coating will be investigated. In addition, an experimental observation of the dispersive behavior, as a function of wavenumber amplitude and direction, will be reported.

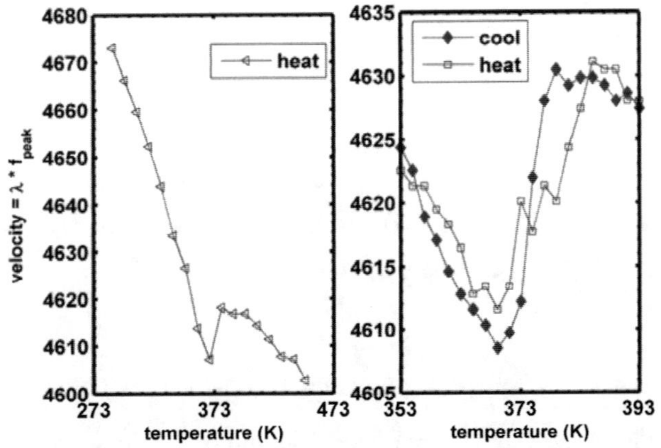

Figure 2. Temperature dependence of the phase velocity of SAW propagating on the surface of the Ni_2MnGa film/MgO substrate sample. The acoustic wavelength $\lambda = 10$ μm. The markers are experimental points, the lines were added to the eye.

Acknowledgements

Authors wish to express their thanks to Anja Backen from IFW Dresden for sample preparation and for kindly providing other characteristics of the tested film. This work was possible with financial support from the KULeuven research project OT/11/064, FWO project KAN2010 1.5.168.10 and the Czech Science Foundation(CSF), project no. P101/12/P428.

References

[1] Sun B, Winey J M, Hemmi N, Dreger Z A, Zimmerman K A, Gupta Y M, Torchinsky D H and Nelson K A 2008 *J. Appl. Phys.* **104** 073517

[2] Verstraeten B, Van Humbeek J, Wevers M, Glorieux C 2013 *Int. J. Thermophys.* DOI 10.1007/s10765-013-1405-3

[3] Sermeus J, Matsuda O, Salenbien R, Verstraeten B, Fivez J and Glorieux C 2012 *Int. J. Thermophys.* **33** 2145–58

[4] Rogers J A and Nelson K A 1994 *J. Appl. Phys.* **75** 1534–56

[5] Salenbien R, Côte R, Goossens J, Limaye P, Labie R and Glorieux C 2011 *J. Appl. Phys.* **109** 093104

[6] Rogers J A, Maznev A A, Banet M J and Nelson K A 2000 *Annu. Rev. Mater. Sci.* **30** 117–57

[7] Kobayashi M, Nakanishi M, Tsujimi Y and Yagi T 2002 *J. Non-Cryst. Solids.* **307-310** 252–6

[8] Straka L, Heczko O, Seiner H, Lanska N, Drahokoupil J, Soroka A, Fähler S, Hänninen H and Sozinov A 2011 *Acta Mater.* **59** 7450–63

[9] Heczko O, Straka L and Seiner H 2013 *Acta Mater.* **61** 622–31

[10] Backen A, Yeduru S R, Diestel A, Schultz L, Kohl M and Fähler S 2012 *Adv. Eng. Mat.* **14** 696–709

[11] Thomas M, Heczko O, Buschbeck J, Rössler U K, McCord J, Scheerbaum N, Schultz L and Fähler S 2008 *New J. Phys.* **10** 023040

[12] Chernenko V A 1999 *Scr. Mater.* **40** 523–7

[13] Backen A 2013 Private communication

[14] Srivastava V K and Chatterjee R 2005 *IEEE Trans. Magn.* **41** 3446–7

[15] Sokolov A, Zhang L, Dubenko I, Samanta T, Stadler S and Ali N 2013 *Appl. Phys. Lett.* **102** 072407

[16] Gusev V, Desmet C, Lauriks W, Glorieux C and Thoen J 1996 *J. Acoust. Soc. Am.* **100** 1514–28

[17] Maznev A A, Nelson K A and Rogers J A 1998 *Opt. Lett.* **23** 1319–21

Journal of Physics: Conference Series **520** (2014) 012014 doi:10.1088/1742-6596/520/1/012014

Automatic Event Detection in Noisy Environment for Material Process Monitoring by Laser AE Method

K Ito[1], H Kuriki[1], H Araki[2], S Kuroda[2] and M Enoki[1]

[1]Department of Materials Engineering, School of Engineering, The University of Tokyo, 7-3-1 Hongo, Bunkyo-ku, Tokyo 113-8656, Japan
[2]High Temperature Materials Unit, National Institute for Materials Science, 1-2-1, Sengen, Tsukuba, Ibaraki 305-0047, Japan

E-mail: ito@rme.mm.t.u-tokyo.ac.jp

Abstract. Laser acoustic emission (AE) method is a unique *in-situ* and non-contact non-destructive evaluation (NDE) method. It has a capability to detect signals generated from crack generation and propagation, friction and other physical phenomena in materials even in high temperature environment. However, laser AE system has lower signal-to-noise ratio compared to the conventional AE system using PZT sensors, so it is difficult to apply this method in noisy environment. A novel AE measurement system to detect events in such difficult environments was developed. This system could continuously record all AE waveforms and enable unrestricted post-analyses. Noise reduction filters in frequency domain coupling with a new AE event extraction using multiple threshold values showed a good potential for AE signal processing. This system was successfully applied for crack monitoring of plasma spray deposition process of ceramic coating.

1. Introduction

Non-destructive evaluation (NDE) methods for process monitoring have many advantages, such as precise controlling of processes and detection of accidental failures, *etc.* In this field, AE method is one of important candidates because it provides *in-situ* monitoring, which can detect various physical events in solid materials *e.g.* crack generation and propagation. Piezoelectric zirconate titanate (PZT) sensors are generally used as AE sensors, however they have applicable temperature limit up to their Curie point. Laser AE method is one choice for high-temperature monitoring. Since laser interferometers are used as AE sensors, non-contact measurement becomes available. However, there still has a problem about signal-to-noise ratio due to noises generated during process monitoring. Their sources and intensities change frequently along a progress of processes. Recently, a novel system called, "Continuous Wave Memory" (CWM), has been developed for detection and analysis of AE waveforms. The CWM system has a capability to continuously record all AE waveforms during the whole measurement time and this makes post-analysis of AE waveforms flexible such as re-configuration of noise filters and threshold voltages.

Thermal barrier coatings (TBCs) with yttria-stabilized zirconia (YSZ) top coat deposited by atmospheric plasma spraying (APS) have been widely used to protect metal substrate from heat and wear damages. In this top coat, vertical cracks were induced during deposition process to improve performance and durability of TBCs [1]. For the sake of process controlling, it is important to monitor the crack generation during APS deposition, however, there are few methods applicable for *in-situ* monitoring. Our previous study has shown that the deposition of Al_2O_3 top coat was successfully monitored by laser AE method [2]. However, AE monitoring during deposition of YSZ top coat was more difficult because AE events generated in YSZ top coat are less numbers and smaller amplitudes than those in Al_2O_3 top coat due to higher toughness values of YSZ. In this study, we applied laser AE method to investigate crack generation during APS of YSZ top coat. An improvement of signal processing based on multiple-threshold method for AE event extraction was carried out and the potential of these methods were discussed.

2. Experimental Setup

Figure 1 shows the experimental setup of a specimen and experimental equipment. The substrate of the specimen was an Inconel® 601 (Nilaco Corp.) disk with a diameter and a thickness of 30 and 5 mm, respectively. The top surface of the substrate was grid blasted to increase the adhesion between the substrate and the bond coat while the bottom surface was mirror-finishing polished to improve the reflectance of a laser beam for better detectability of laser AE. CoNiCrAlY particles (AMDRY 9954, Sulzer Metco Ltd.) were preliminary deposited as a bond coat by the high velocity oxygen fuel (HVOF) spray method. Then, the prepared specimen was fixed to the internal jig as shown in figure 1(b). The external and internal jigs were separated and vibration insulator was attached. The jigs were carefully installed to avoid superfluous vibration and laser scattering by suspended particles. A plasma torch (SG-100, Praxair, Inc.) scanned over the specimen with a speed of 50 mm/s (figure 1(c)). YSZ particles (K-90, Showa Denko K. K.) were sprayed through a hole on the external jig with a diameter of 40 mm just over the specimen. Temperatures were also measured by type K thermocouples with a frequency of 100 SPS (samples/s).

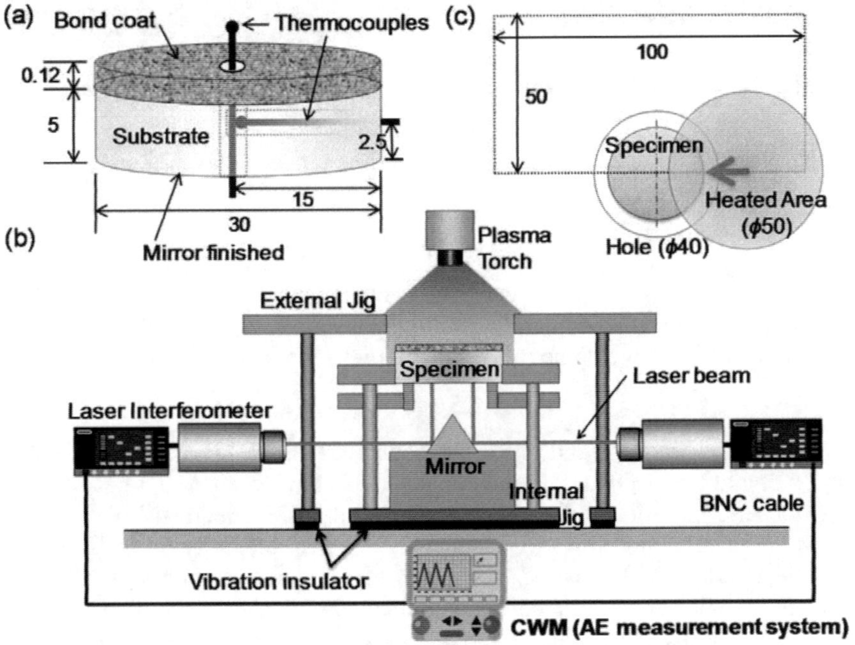

Figure 1. Experimental setup for laser AE monitoring of APS process.
(a) Specimen, (b) Equipments, (c) Scanning path of the plasma torch.

Four laser interferometer units (AT3600S and AT0022, Graphtec Corp.) were used as AE sensors. All laser beams were focused on the bottom surface of the specimen to monitor the speed of out-of-plane vibration (figure 1(b)). The output signals of the laser interferometers were continuously recorded by CWM with a frequency of 10 MSPS. Sampling resolution and range were 12 bit and ± 5

V, respectively. Synchronized analog-to-digital converter boards (PCI-3525, Interface Corp.) and parallelized hard disk drives (HDDs) were used in CWM for this sampling.

3. Results and Discussion

Figure 2(a) show an example of short time Fourier transform (STFT) of detected signal. It was evident that STFT is helpful method to separate valid signals from noise by their frequency characteristics. The noise filter was applied by following equation [3, 4]

$$I_1(t,f) = \max\{0, I_0(t,f) - \lambda(f)\} \tag{1}$$

where t is time, f is frequency, I_o and I_1 are the intensity matrix before and after the filtering and $\lambda(f)$ is a subtraction value for noise reduction (NR). When $\lambda(f)$ is sufficiently large in some frequency band, the above equation behaves as a frequency filter. It also behaves as a background noise filter when $\lambda(f)$ is an appropriate value in entire frequency band. By apply this filter with $\lambda(f) = \infty$ for $f < 100$ kHz and $\lambda(f) = 3000$ a.u. for $f \geq 100$ kHz, the noise as shown in figure 2(a) could be removed and the result is shown in figure 2(b). Since noises fluctuated over the monitoring period, continuous recording enabled post-configuration for optimization of NR filter based on confirmation of frequency characteristic of actual waveform. After NR, the intensity matrix was converted into the waveform for further analysis by inverse STFT method.

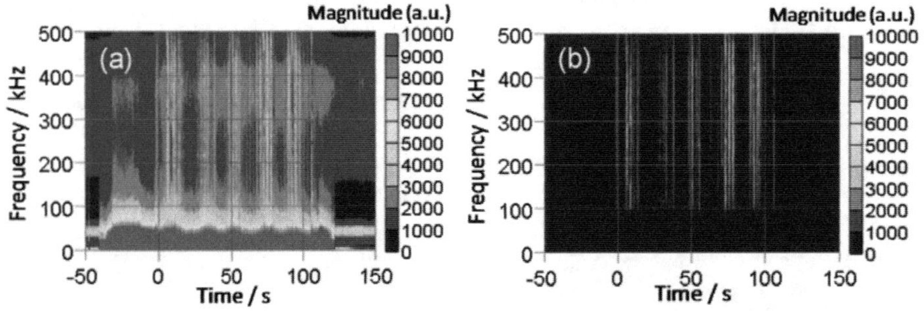

Figure 2. Time-frequency-intensity matrix (a) before and (b) after the NR filter.

Figure 3 shows a schematic of hit detection method in this study. In general, start and end time of an AE hit are decided by AE count, *i.e.* threshold crossing. However, decision of the last count is difficult if the waveform is noisy or hits occur frequently. Therefore, false recognitions might be occurred, *e.g.* a long duration hit is recognized as multiple hits and/or a series of short duration hits are recognized as one unified hit. To overcome this problem, the end of an AE hit is determined as

$$\Delta t_s \geq a \cdot \min\{\Delta t_{\mathrm{Max}}, \max(\Delta t_{\mathrm{Min}}, \Delta t_d)\} \tag{2}$$

where Δt_s is silent time (time from the end of the last count), a is a pre-set ratio, Δt_d is duration of a hit (time between the first and the last count), Δt_{Min} and Δt_{Max} are limiters of Δt_s. When the last AE count is identified after Δt_d period the first count, the end of this hit will be decided if there is no AE count until $a\Delta t_d$ period from the last count. If an AE count is detected in this time, Δt_d is extended to the new last AE count and the judge will be restarted. The value of Δt_d increases with lower threshold voltage (V_{th}). However, when V_{th} is lower than the noise level, inequality (2) is not satisfied and no AE hit will be detected (figure 3(b)). In this study, the default values of a, Δt_{Min} and Δt_{Max} were set as 0.5, 100 μs and 10 ms, respectively. The selection of suitable V_{th} is very important for AE extraction. Generally, a low fixed value of V_{th} is preferred when the noise level is sufficiently low and stable, however, this is not practicable for fluctuated-noise environment. In this study, multiple-threshold method was used for AE hit and event detection

Figure 4 showed an example of AE events detection during some part of an experiment by multiple-threshold method (50, 80, 100 and 150 mV). It is noted that events shown in this figure were simultaneously detected from at least three channels. No events were detected with $V_{\mathrm{th}} = 50$ mV

Journal of Physics: Conference Series **520** (2014) 012014 doi:10.1088/1742-6596/520/1/012014

because this V_{th} was always lower than the noise level. The numbers of events with V_{th} of 80, 100 and 150 mV were two, three and seven, respectively. The V_{th} of 80 mV was available only when the noise level was relatively low. In contrast, AE events with small amplitudes were discarded with V_{th} of 150 mV even if the noise level was sufficiently low. It is clear that the number of detectable AE events decreased by using one fixed V_{th}. The multiple-threshold method showed a potential to extract more detectable AE events by summation of all detected events from each V_{th}. This method also provided an automatic AE event detection without manual adjustment of V_{th}.

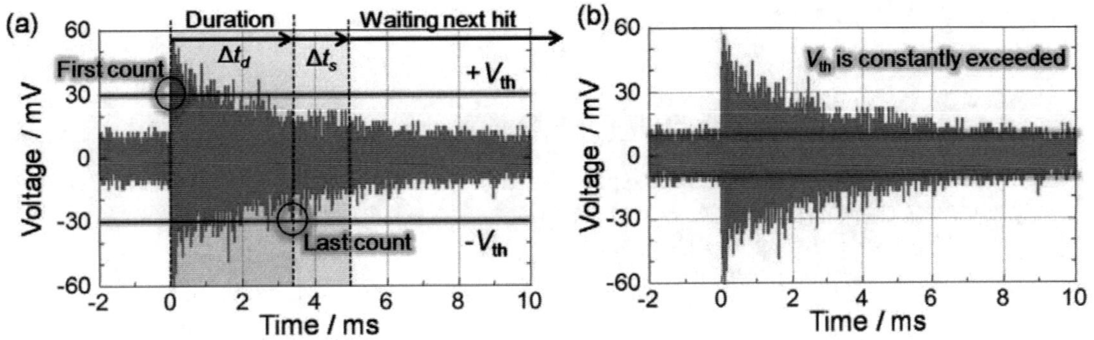

Figure 3. Schematic of AE hit detection with (a) high and (b) low threshold voltages.

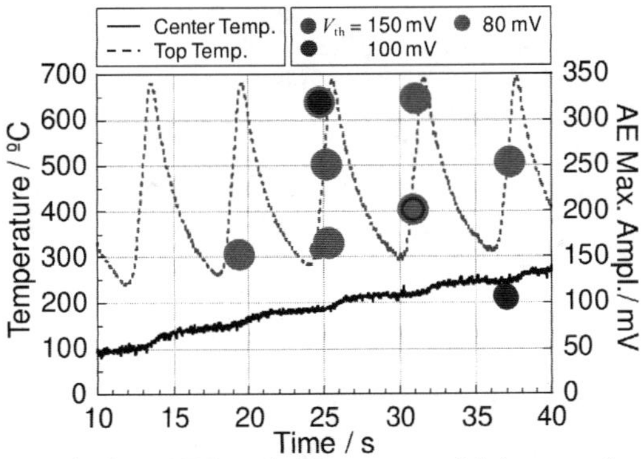

Figure 4. Temperature history during APS and AE events which were detected by multiple-threshold method.

4. Conclusions

Continuous waveform recording enabled the flexible and strong filters for NR in frequency domain. A combination of frequency filter and background noise filter was effective to improve the signal-to-noise ratio. Multiple-threshold method for AE hit and event detection made it possible to follow fluctuating noise level during a noisy manufacturing process of materials such as plasma spray.

Acknowledgement

We thank to Mr. Hiraoka and Mr. Komatsu for supporting us. This work was partially supported by MEXT/JSPS KAKENHI Grant Number 23246124.

References

[1] (For example) Karger M, Vaßen R, Stöver D 2011 *Surf. Coat. Tech.* **206** 16
[2] Ito K, Ohmata S, Kobayashi K, Watanabe M, Kuroda S, Enoki M 2010 *Mater. Trans.* **51** 1272
[3] Abbate A, Koay J, Frankel J, Schroeder S C, Das P 1997 *IEEE Trans. Ultrason., Ferroelect., Freq. Contr.* **44** 14
[4] Donoho D L 1995 *IEEE Trans. Info. Theory* **41** 613

Journal of Physics: Conference Series **520** (2014) 012015 doi:10.1088/1742-6596/520/1/012015

Application of Laser Ultrasound to Noncontact Temperature Profiling of a Heated Hollow Cylinder

A Kosugi[1], Y Ono[1], I Matsuya[2] and I Ihara[2]
[1]Graduate School of Nagaoka University of Technology,
Nagaoka, Niigata 940-2188, Japan
[2]Department of Mechanical Engineering, Nagaoka University of Technology,
Nagaoka, Niigata 940-2188, Japan

E-mail: ihara@mech.nagaokaut.ac.jp

Abstract. A new ultrasonic method to monitor the temperature distributions of a thick-walled hollow cylinder whose inner surface is heated is presented. This method basically consists of laser ultrasonic measurements and a heat conduction analysis of the heated hollow cylinder. Both longitudinal wave (LW) and surface acoustic wave (SAW) are used for estimating the temperature distribution of the cylinder. To demonstrate the validity of the proposed method, a numerical simulation and an experiment are carried out. In the numerical simulation, a steel hollow cylinder (the inner and outer diameters are 20 mm and 100 mm, respectively) is uniformly heated by a constant heat flux at the inner surface. It has been verified in the simulation that temperature distributions estimated by the proposed method completely agreed with the theoretical results. In the experiment, an aluminum hollow cylinder (the inner and outer diameters are 20 mm and 50 mm, respectively) is heated by pouring molten metal at 300 °C into the hollow cavity of the cylinder. A laser ultrasonic system is used for measuring both LW and SAW of the heated cylinder. It has also been demonstrated that the proposed method can measure the change in the temperature distribution during heating.

1. Introduction

There is an increasing demand for temperature measurements of heated materials in the fields of engineering and industry. In particular, it is urgently required to monitor and control internal temperatures of hollow pipes or vessels filled with high temperature fluids or gases because the temperature state of such pipes is directly related to serious damage of pipes due to chemical reactions, corrosion, erosion, thermal stress and thermal fatigue. It is, therefore, imperative to develop an accurate in-situ monitoring technique of internal temperature distributions of heated pipes or vessels.

Ultrasound, owing to its high sensitivity to temperature, is expected to be a promising candidate for measuring the temperature of materials. Because of the advantages of ultrasonic measurements such as non-invasiveness and faster response, some works on the application of ultrasound to temperature estimation have been carried out extensively [1]-[6]. In our previous works [7]-[9], an effective methods consisting of an ultrasonic time-of-flight measurement and a heat conduction equation were developed for measuring temperature distributions of heated materials. The advantage of this method is that there is no need for the thermal boundary conditions at the heating surface where the thermal state is often unstable and difficult to know properly.

Journal of Physics: Conference Series **520** (2014) 012015 doi:10.1088/1742-6596/520/1/012015

In this work, based on our previous methods [7]-[9], a method with a laser ultrasonic technique for monitoring the temperature distributions of a thick-walled hollow cylinder has been developed. To demonstrate the feasibility of this method, numerical and experimental demonstrations are carried out.

2. Determining temperature distribution of a hollow cylinder

The principle of temperature determination by ultrasound is based on the temperature dependence of ultrasonic wave velocity. Assuming that a hollow cylinder is uniformly heated at the inner or outer surface and has a one-dimensional temperature distribution in the radial direction as shown in figure 1, transit times for a longitudinal wave (LW) propagating along the radial direction and a surface acoustic wave (SAW) propagating along the outer surface of the cylinder are given by

$$t_{LW} = \int_0^{L_{LW}} \frac{1}{v_{LW}(T(x))} dx \tag{1}$$

$$t_{SAW} = \frac{L_{SAW}}{v_{SAW}(T)} \tag{2}$$

where L is the propagation distance of the ultrasonic wave and v is the ultrasonic velocity which is a function of temperature T. It should be noted that the LW propagates along the radial direction and the SAW propagates along the outer surface. The temperature at the outer surface is simply determined from SAW using equation (2). The internal temperature distribution in the radial direction is then determined by solving a one-dimensional heat conduction equation in a cylindrical coordinate system [10]. To determine the temperature at the inner surface from equations (1) and (2), an ultrasonic time-of-flight measurement and a finite difference calculation to solve the heat conduction equation in the cylindrical coordinate system are effectively combined [7]-[9]. The inner surface temperature is then determined from the transit time of the LW and both surface and internal temperatures obtained precedently. The advantage of this method is that there is no need to know the thermal boundary conditions at not only outer surface but also inner surface at where the boundary condition is usually unknown. It is also noted that the present method can be applied to the temperature estimation of a material having an unsteady temperature distribution.

3. Numerical demonstration

To verify the validity of the proposed method, a numerical simulation for the ultrasonic temperature estimation is carried out. A steel hollow cylinder with an axisymmetric temperature distribution as shown in figure 1 is used as the simulation model. The cylinder with inner and outer diameters of 20 mm and 100 mm, respectively, is heated by a constant heat flux (500 kW/m^2) at the inner surface. At first, the temperature distribution of the cylinder and its variation during heating are theoretically obtained by solving the heat conduction equation [10]. From these results, the transit times of ultrasonic waves for LW and SAW are determined from the equation (1) and (2). Using the transit times, the temperature distributions are then determined from the procedure mentioned above (the proposed method). Both temperature distributions estimated by the proposed method and estimated theoretically with the heat conduction equation are compared with each other. In the numerical demonstrations, the thermal diffusivity coefficient α = 7.53 mm^2/s, time interval for the finite

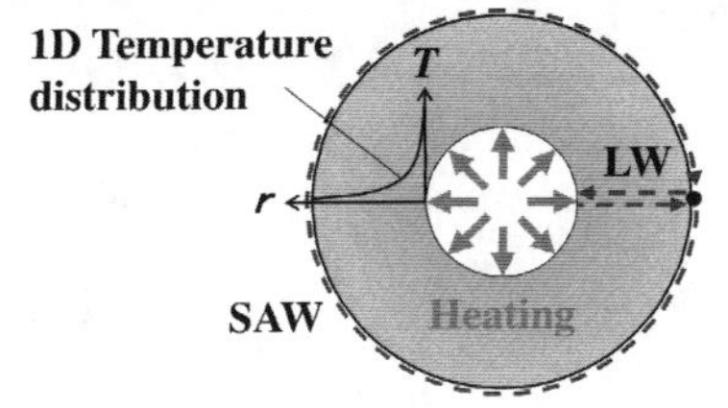

Figure 1. Schematic of an axisymmetric temperature distribution of an internally heated hollow cylinder and the LW and SAW propagations.

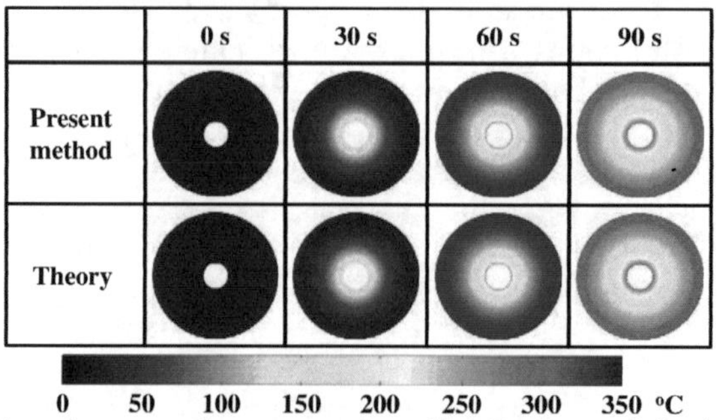

Figure 2. Comparison between the temperature distributions of the hollow cylinder estimated by the ultrasonic method and theoretical calculation.

difference calculation τ = 0.01 s, spatial interval for the finite difference calculation h = 1.0 mm, temperature dependence of the LW velocity v_{LW}=-0.763T+5736.2 m/s, and temperature dependence of the SAW velocity v_{SAW}=-0.538T+2895.9 m/s are used.

Figure 2 shows the estimated temperature distributions and their variations with the elapsed time after the heating starts. It can be seen that the temperature distribution in the cylinder changes gradually with the elapsed time. It is noted that the ultrasonically estimated results by the proposed method (upper) agree with the theoretically estimated results by the heat conduction equation (lower) completely. Thus, it is verified that the proposed ultrasonic method does work properly.

4. Experimental demonstration

Based on the numerical demonstration, the feasibility of the proposed method is demonstrated experimentally. Figure 3 shows the schematic of the experimental setup and the cross section of the cylinder used in the experiment. An aluminum hollow cylinder with inner and outer diameters of 20 mm and 50 mm, respectively, is used as the specimen and is uniformly heated by pouring molten metal at 300 °C into the hollow cavity of the cylinder. It should be noted that temperature distributions cannot be determined by numerical analyses because of lack of thermal boundary conditions. A laser ultrasonic system consisting of a pulsed laser generator (Nd:YAG, wavelength 1064 nm, energy 180 mJ, pulse width 3 ns) and a laser interferometer (Nd:YAG, wavelength 532 nm, power 200 mW) is used for measuring both LW and SAW. The two waves are continuously acquired every 0.033 s with a PC based acquisition system. The sampling rate of ultrasonic signals is 100 MHz. The transit time of each wave and its variation during heating is calculated by the cross correlation method. Using the transit times and their variations with the elapsed time, the temperatures at inner and outer surfaces and the temperature distributions of the cylinder are estimated. To compare with the ultrasonically estimated results, thermocouples are inserted into the cylinder at 5 and 10 mm from the inner surface

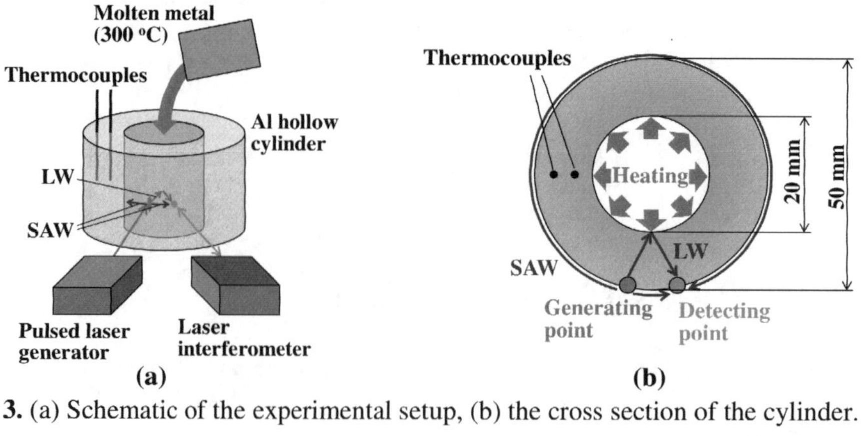

Figure 3. (a) Schematic of the experimental setup, (b) the cross section of the cylinder.

and used for internal temperature measurements. In the ultrasonic estimation, the thermal diffusivity coefficient α = 49.2 mm^2/s, temperature dependence of the LW velocity v_{LW}=-0.9797T+6428.9 m/s, and temperature dependence of the SAW velocity v_{SAW}=-0.8465T+2984.2 m/s are used.

Figure 4 shows variations in the temperature distributions of the cylinder with the elapsed time after heating starts, where the ultrasonically estimated results (line) are compared with those measured using the thermocouples. It can be seen that temperature distributions are gradually changed with the elapsed time. Although the ultrasonic results almost agree with those measured using the thermocouples, there are certain discrepancies between them in the early stage of heating such as within a few seconds after heating starts. This is likely because of relatively slow time response of the conventional thermocouples.

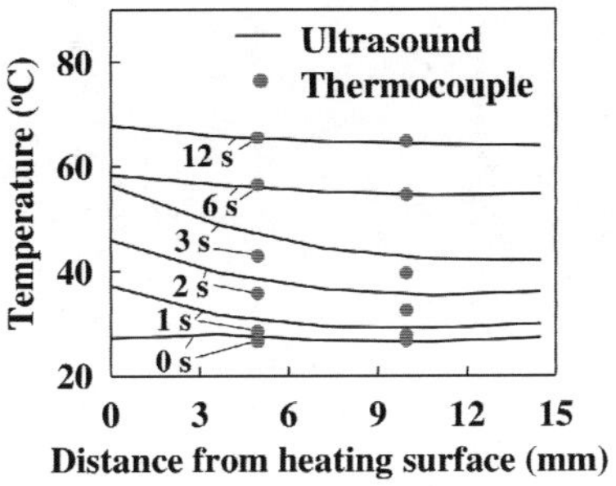

Figure 4. Variations in temperature distributions of the hollow cylinder with elapsed time, determined by the ultrasonic method (line) and measured using thermocouples (circle).

5. Conclusion

A noncontact ultrasonic method with a laser ultrasonic technique for measuring internal temperature distributions of a thick-walled hollow cylinder whose inner surface is heated, is presented. The validity of the proposed method is verified through a numerical simulation and its practical feasibility is also demonstrated experimentally. Although further study is necessary to improve the accuracy, practicability and robustness in the method, it is highly expected that the ultrasonic method could be a useful means for on-line temperature monitoring of hollow pipes or vessels filled with high temperature fluids or gases.

Acknowledgment

Supports from the JSPS Grant-in-Aid for Scientific Research (B) 25289238 and Toyota Motor Co. are greatly appreciated.

References

[1] Degertekin F L, Pei J, KhuriYakub B T and Saraswat K C 1994 *Appl. Phys. Lett.* **64** 1338-40
[2] Simon C, VanBaren P and Ebbini E S 1998 *IEEE Trans. Ultrason. Ferroelectr. Freq. Control* **45** 1088-99
[3] Balasubramaniam K, Shah V V, Costley R D, Boudreaux G and Singh J P 1999 *Rev. Sci. Instrum.* **70** 4618-23
[4] Matsumoto T, Nagata Y, Nose T and Kawashima K 2001 Rev.Sci. Instrum. 72 2777-83
[5] Huang K N, Huang C F, Li Y C and Young M S 2002 *Rev. Sci. Instrum.* **73** 4022-27
[6] Mizutani K, Kawabe S, Saito I and Masuyama H 2006 *Jpn. J. Appl. Phys.* **45** 4516-20
[7] Takahashi M and Ihara I 2009 *Jpn. J. Appl. Phys.* **48** 07GB04
[8] Ihara I and Tomomatsu T 2011 *IOP Conf. Ser.: Mater. Sci. Eng.* **18** 022008
[9] Yamada H, Kosugi A and Ihara I 2011 *Jpn. J. Appl. Phys.* **50** 07HC06
[10] Ozosik M N1968 *Boundary Value Problems of Heat Conduction* (Scranton: International Textbook Company) pp 412-420

Journal of Physics: Conference Series **520** (2014) 012016 doi:10.1088/1742-6596/520/1/012016

Application of Air-Coupled Ultrasound to Noncontact Evaluation of Paper Surface Roughness

M N F Saniman[1] and I Ihara[2]

[1] Graduate School of Nagaoka University of Technology, 1603-1 Kamitomioka, Nagaoka, Niigata 940-2188 Japan
[2] Mechanical Engineering Department, Nagaoka University of Technology, 1603-1 Kamitomioka, Nagaoka, Niigata 940-2188 Japan

E-mail: moll@stn.nagaokaut.ac.jp

Abstract. An approach for characterizing paper surface quality by ultrasound as an alternative non-contact method is presented. In this work, an air-coupled ultrasound at frequency range from 0.3 MHz to 4.2 MHz has been applied to surface roughness characterization, where a series of sandpapers and pure papers having random and relatively wide range of root-mean-square of roughness Rq from 2.0 to 92.8 μm are employed as specimens. The amplitude of reflected wave from each specimen is measured with pulse-echo configuration at normal incidence. A Kirchhoff-based scattering model is used to express the scattering phenomena from random rough surfaces and the relations between the normalized amplitude of the reflected wave and surface roughness parameters are then examined. It has been shown through the experiments that high frequency air-coupled ultrasound up to 4 MHz is useful to characterize surface roughness in the order of few microns of Rq. In addition, it has been suggested that an irregularity of paper surface geometry such as skewness could be characterized from the deviation of the normalized amplitude.

1. Introduction

Paper is an important material for storing information either by illustrations or written words. Print defects such as missing dots, ink bleeding and print mottle are usually associated with the printability of paper [1]. The surface roughness of paper is one of the most important factors influencing the printability [2].

Traditionally, the surface roughness of papers is often evaluated by air leak method [1]. This method is, however, not suitable for on-line use because the probe has to be contacted with paper surface in this method. As one of the alternatives for evaluating paper surface roughness, air-coupled ultrasound is considered to be a promising candidate because it is non-destructive, non-contact, convenient, fast, and sensitive to even soft matters such as cellulose fibre based materials e.g. papers and woods [3]. In addition, a wide frequency range of ultrasound may allow evaluating a wide range of surface roughness [4].

Although there are several works about the application of air-coupled ultrasound on paper characterization [4-7], a relatively lower frequency range less than 1 MHz is used in those works and paper roughness evaluations have not been sufficiently studied. It is expected to characterize surface roughness in the range of a few microns if a higher frequency larger than 1 MHz is used in air-coupled ultrasound measurements. In this work, an air-coupled ultrasound at frequency range from 0.3 MHz to 4.2 MHz has been applied to surface roughness characterization of a series of sandpapers. The feasibility of the use of such high frequency air-coupled ultrasounds for characterizing surface roughness of pure papers is then examined.

2. Theory

2.1. Kirchhoff's scattering model

When an elastic wave is reflected from a rough surface, the reflected fields consist of a specularly reflected coherent and diffusely reflected incoherent component depending on the degree of roughness. Kirchhoff theory is employed to understand the relation between surface roughness and reflected wave intensity. For a plane wave incidence, the overall scattered field reflected from a Gaussian distributed rough surface, I, can be written as [8]

$$I = I_{coh} + I_{incoh} \tag{1}$$

where I_{coh} *and* I_{incoh} are the intensity of coherent and incoherent components, respectively. At normal specular angle (incident angle, θ_i = reflected angle, θ_r = 0°), the intensity of coherent component I_{coh} is given by

$$I_{coh} = I_0 \exp(-2k^2 Rq^2) \tag{2}$$

where I_0 and k are the reflected wave intensity from a perfect smooth surface and the wavenumber, respectively, and the root-mean-square of roughness, Rq, is determined from [9]

$$Rq = \sqrt{\frac{1}{N}\sum_{i=1}^{N} h_i^2} \tag{3}$$

where h and N are the height deviation from the mean value of the surface height profile and the number of data points, respectively.

2.2. Skewness

Skewness, Rsk, is the measure of asymmetry of the height profile distribution and is given by [9]

$$Rsk = \frac{1}{NRq^3}\sum_{i=1}^{N} h_i^3 \tag{4}$$

Zero skewness means the height profile is symmetrically distributed and such profile is known as a Gaussian distribution i.e. a profile that has as many peaks as valleys. Positive skewness means the profile has a lot of high peaks while negative skewness means that the profile has a lot of deep valleys as shown in figure 1.

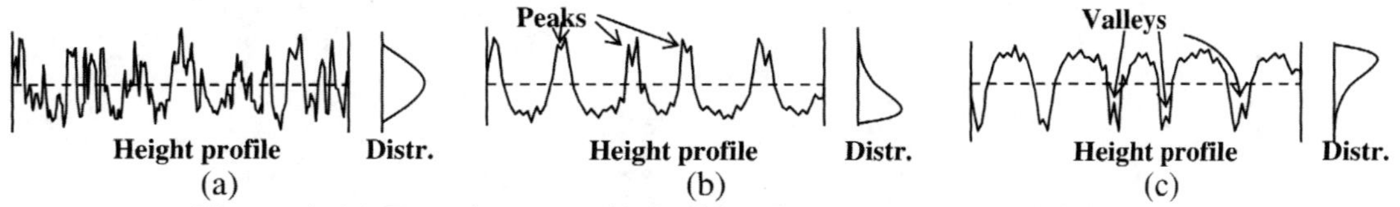

Figure 1. (a) Zero skewness (b) Positive skewness (c) Negative skewness.

3. Experimental setup

Three types of air-coupled ultrasonic transducers with frequencies 0.4, 2 and 4 MHz from Japan Probe Co. are used in a pulse echo configuration with normal incident to a specimen surface at the specimen/transducer distance of 30, 17 and 3 mm, respectively. Each transducer is driven using a pulser/receiver BPR-4000 from Ritec Co. To reduce the measurement noise of the signals, the average values of the reflected amplitudes at different locations of each specimen are employed for evaluation. The specimens used in this study are a series of sandpapers having a relatively large range of Rq (2.7 ~ 92.8 μm) and several kinds of pure papers having a small range of Rq (3.0 ~ 6.1 μm), where the roughness values are measured by a stylus profiler according to ISO 4288:1996 specifications. In addition, a smooth acrylic plate (Rq = 0.02 μm) is used as a reference surface for normalization.

4. Results and discussions

Figure 2(a) shows the measured results showing the relationship between the reflected amplitude of ultrasound and surface roughness for a series of sandpapers having Gaussian height distribution. It is noted that the roughness Rq is normalized by multiplying with frequency f, to adequately examine the variations of the normalized amplitude at any frequency. It can be seen that the measured amplitudes agree well with the theoretical results from equation (2) illustrated by a solid line, for all frequency components. The normalized amplitudes in figure 2(a) are used to ultrasonically estimate the Rq values for each specimen by using equation (2) and then compared with those measured by a stylus profiler as shown in Figure 2(b). It is found that both results agree well with each other within ±10% errors in the frequency range approximately 5 to 60 μm•MHz. Based on this range, it is possible to make proper selections for the operating frequency and the corresponding measurable range of Rq for performing an appropriate ultrasonic roughness estimation. For $Rq{\cdot}f$ less than 5 μm•MHz, estimation errors are caused due to the accuracy in measuring the amplitude using the present measurement system. For $Rq{\cdot}f$ larger than 60 μm•MHz, the errors may be caused due to the generation of the incoherent components in the reflected waves.

It has been found in Figure 2(b) that a higher frequency such as 4 MHz is appropriate for evaluating relatively small values of Rq, e.g. 2 to 6 μm, such as those for pure papers. Figure 3(a) shows the measured amplitudes by 4 MHz air-coupled ultrasound for pure paper surfaces having relatively small values of roughness. As we can see, the normalized amplitudes of each paper almost agree with the theoretical ones except for a few specimens having small Rq. Detailed investigations with the statistical descriptions of height profiles of each specimen measured by a stylus profiler are conducted. It is found that the normalized amplitudes reflected from paper surfaces with Gaussian distributed height profile, as shown in figure 3(c), are in good agreements with the theoretical ones while those that reflected from the surfaces having negatively skewed height profiles, as shown in figure 3(b), deviate markedly from the theoretical ones. Negatively skewed height profile means that

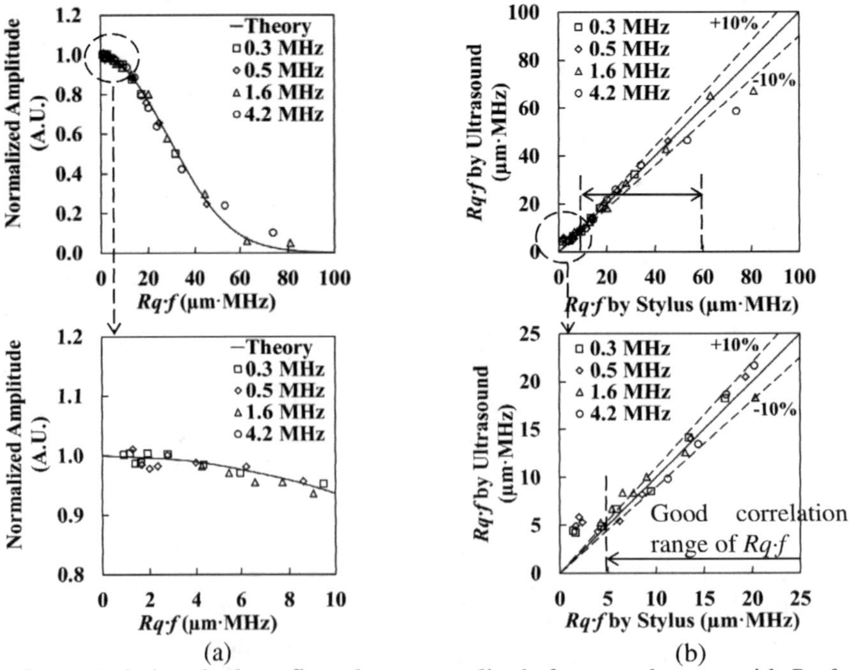

Figure 2. (a) Variations in the reflected wave amplitude from sandpapers with $Rq{\cdot}f$, (b) Comparison between the roughness estimated by a stylus and ultrasonic methods.

the paper specimen has a lot of deep valleys with small height variations at the reflection surface as shown in figure 3(b). This kind of surface is formed during the hard beating in a paper making process. The results shown in figure 3 suggest that paper surface skewness may be possible to be evaluated by a high frequency air-coupled ultrasound.

The measurement can be improved further by improving the ultrasonic beam profile to be more Gaussian and applying advanced signal processing methods. Although the ultrasonic response as a function of incidence angle is of interest, theoretical calculations show that normal incidence angle is the most sensitive to any roughness changes and it have been shown experimentally that the selection of incidence angle has no significant effect on the overall measurement results [10].

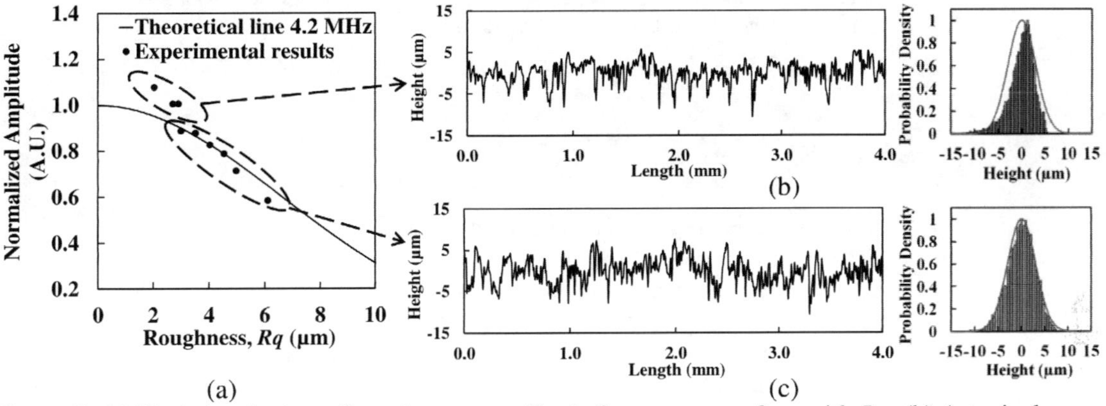

Figure 3. (a) Variation in the reflected wave amplitude from paper surface with *Rq*. (b) A typical height profile of a negatively skewed surface and its height distribution. (c) A typical height profile of a Gaussian distributed surface and its height distribution.

5. Conclusions

An air-coupled ultrasound at frequency range from 0.3 MHz to 4.2 MHz has been applied to a series of sandpapers and pure papers having random and relatively wide range of roughness *Rq* from 2.0 to 92.8 μm, to examine the feasibility of characterizing paper surface roughness. It has been shown through the experiments that high frequency air-coupled ultrasound up to 4 MHz can be a useful means to characterize surface roughness in the order of a few microns of *Rq*. In addition, it is suggested that the air-coupled ultrasound has the potential to characterize an irregularity of paper surface geometry such as skewness. The quantitative evaluation of skewness will be conducted in a future work.

References

[1] Fellers C *et al* 2009 *Pulp and Paper Chemistry and Technology: Paper Products, Physics and Technology* vol 4, ed Ek M *et al* (Berlin: Walter de Gruyter) chapter 8 pp 213-23
[2] Popil R E 1996 *Proc. TAPPI Process & Prod. Qual. Conf.* (Cincinnati: TAPPI Press) pp 103-8
[3] Žukauskas E, Raišutis R, Voleišis A, Vladišauskas A and Kairiūkštis L 2008 *Ultrasound* **63** p 33
[4] Sukmana D D and Ihara I 2007 *Jap. J. Appl. Phys* **46** pp 4508-13
[5] Santos M, Perdigao J and Velho J 2007 *Insight* **49** pp 146-50
[6] Kazys B and Stolpe P 2001 *NDT&E International* **34** p 259
[7] Habeger C, Mann R and Baum G 1979 *Ultrasonics* **17** pp 57-62
[8] Ogilvy J A 1991 *Theory Of Wave Scattering From Random Rough Surfaces* (Bristol: IOP Publishing) chapter 4 p 88
[9] Gadelmawla E S, Koura M M, Maksoud T M A, Elewa I M and Soliman H H 2002 *J. Mater. Process. Technol.* **123** pp 133-45
[10] Sukmana D D and Ihara I 2005 *Jap. J. Appl. Phys.* **44** pp 4517-20

Journal of Physics: Conference Series **520** (2014) 012017 doi:10.1088/1742-6596/520/1/012017

Orientation Characterisation of Aerospace Materials by Spatially Resolved Acoustic Spectroscopy

Wenqi LI[1], Jethro COULSON[1], John W AVESON[2], Richard J SMITH[1], Matt CLARK[1], Michael G SOMEKH[1], Steve D SHARPLES[1]

[1] Applied Optics Group, Electrical Systems and Optics Research Division, Faculty of Engineering, University Park, University of Nottingham, NG7 2RD, UK;

[2] Department of Materials Science and Metallurgy, University of Cambridge, 27 Charles Babbage Road, Cambridge, CB3 0FS, UK

E-mail: wenqi.li@nottingham.ac.uk

Abstract. Material characteristics in metals such as strength, stiffness and fracture resistance are strongly related to the underlying microstructure. The crystallographic structure and orientation are related to the ultrasonic properties through the stiffness matrix. In individual grains it is possible to analytically determine the ultrasonic velocity from the orientation and stiffness, or determine the stiffness from the known orientation and measured velocity. In this paper we present a technique for imaging the crystallographic orientation of grains in metals using spatially resolved acoustic spectroscopy (SRAS) and a novel inverse solver that can determine the crystallographic orientation from the known stiffness matrix for the material and the SRAS velocity measurement. Previously we have shown the ability of this technique to determine the orientation on single crystal nickel samples; we extended the technique to multigrain industrial metals, such as aluminium, nickel and Inconel. The comparison between SRAS and electron backscatter diffraction (EBSD) on the nickel sample is presented. SRAS is a fast, accurate, quantitative and robust technique for imaging material microstructure and orientation over a wide range of scales and industrial materials.

1. Introduction

SRAS — spatially resolved acoustic spectroscopy — is a nondestructive laser ultrasound method which maps the surface acoustic wave (SAW) velocity of a material surface. The velocity contrast of SAWs reveal the material microstructure and orientation of crystals. Crystal orientation can also be evaluated by other methods such as electron backscatter diffraction (EBSD), or in principle, by scanning acoustic microscopy [1] by analysing the phase shift of a narrow band SAW signal. Although the spatial resolution of SRAS can not compete with EBSD, SRAS has advantages such as being non-contact, having no restriction on sample size and easier surface preparation requirements. The present paper extends previous work to show how it may be used to obtain quantitative crystallographic orientation.

A search algorithm for solving the inverse problem of finding the crystal orientation from the velocity measurements has been developed. Crystallographic orientation images obtained from SRAS using this algorithm are presented as inverse pole figures (IPF) for Inconel and aluminium samples. IPFs for a nickel sample are presented derived from both SRAS and EBSD data.

Journal of Physics: Conference Series **520** (2014) 012017 doi:10.1088/1742-6596/520/1/012017

2. SRAS

SRAS can be used to obtain SAW velocity information on the local area where the waves are generated. The SAW is generated by projecting a grating pattern of laser light on the sample surface. At the sample, typical pulse energies are 12.8μJ per 1.2ns pulse at a repetition rate of 2kHz; ~80mW is used in the detection system at the sample. By varying the frequency by using a broadband (e.g. Q-switched) laser in combination with a fixed grating, the local velocity v therefore calculated through $v = f\lambda$ where λ is the grating period and frequency f is given by doing an FFT on the detected waveform [2, 3]. The spatial resolution is determined by the grating size, generally 100μm for the images presented in this paper.

3. Determining the orientation from SRAS

Determining the velocity from known orientation and elasticity matrix is analytically straight forward, as is obtaining the elasticity matrix from known orientation and velocity. However obtaining the orientation from the measured velocity and elasticity matrix is difficult because it is very ill-conditioned and there is no analytical solution. We have developed a search technique to solve this inverse problem. It finds the orientation by matching the measured velocity at many different propagation directions to the analytically determined velocity from a given elasticity matrix.

3.1. Determining the velocity from the elasticity matrix

Farnell [4] provides a good description of a method to calculate SAW and pseudo-SAW velocities on different crystallographic structures by using iterative search procedures from known material elastic constants. We have made slight modifications to the approach described by Farnell in order to perform the current calculations [2, 5] and a SAW velocity model is calculated. It deserves note that because of the structural symmetry of crystals — for instance the planes (hkl) are mirror symmetrical with (khl) of the cubic crystals — the model includes the planes whose normals are in the area between [001], [011], [101] and [111].

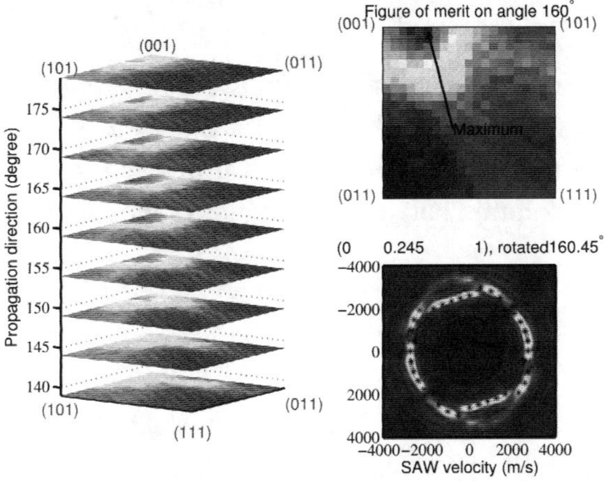

Figure 1: Curve fitting procedure. The figure of merit is on the left as a function of planes (x and y axes) and propagation directions (z axis); the maximum value is denoted by the arrow in the top right figure at the propagation direction of 160°. The bottom right figure displays the output plane on the experimental velocity spectrum; it shows the plane (0 0.245 1) with 160.45° rotation from the reference direction.

3.2. Solving the inverse problem

The challenge is to determine the crystallographic orientation from the measured velocities by reference to the model calculation. The 'raw' measurement of SRAS provides, for each propagation direction angle ϕ, a plot of output signal versus wavelength (or velocity), $A(\phi, v)$, which is presented in right bottom of figure 1. The theoretical model velocities are defined as $v(h,k,l,\theta)$, where the plane is (hkl), and θ is the angle on the plane. The figure of merit is the sum of $A(\phi, v(h,k,l,\theta))$ at each (hkl) and θ calculated according to equation 1

$$F(h,k,l,\theta) = \sum_{\phi=0}^{\phi=\frac{n-1}{n}\pi} A(\phi, v(h,k,l,(\phi-\theta))), \; (n = 1, 2, 3, ...) \quad (1)$$

where n represents the number of evenly spaced velocity measurements used to determine the velocity. Therefore, by repeating the calculation on all planes and all directions, the largest value for F can be found indicating the best match.

4. Experimental validation on multigrain materials

We measure the SAW velocity in multiple directions over an aluminium and an Inconel 617 sample surface by SRAS. IPFs which present the orientation of grains are shown in figure 2. These images show us the high consistency of the SRAS results and the search algorithm performance on a material with low anisotropy (aluminium) even though there are no complementary EBSD results available due to the size of the samples.

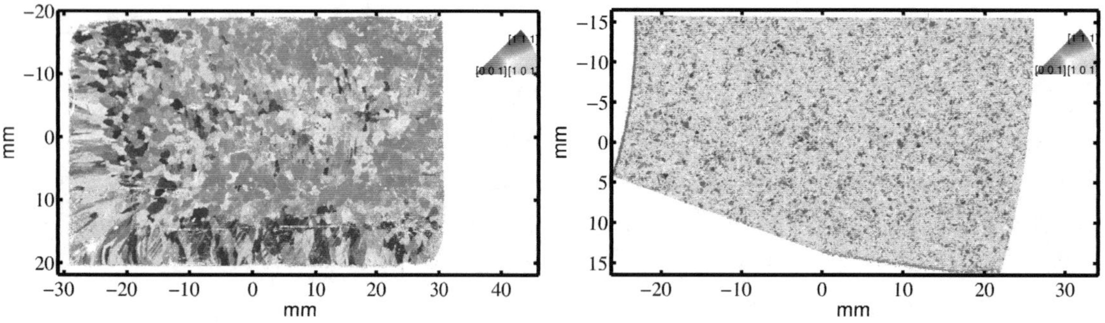

(a) Al: pixel size 100×50μm.

(b) Inconel 617: pixel size 100×50μm.

Figure 2: Crystallographic orientation maps of aluminium and Inconel 617 samples.

Figure 3 demonstrates the orientation maps of a nickel-based superalloy (CMSX4) sample measured by EBSD and SRAS techniques and showing that the SRAS result agrees well with the EBSD data. The resolution of SRAS is not competitive to EBSD but sufficient for the applications which have lower requirements for the resolution but higher demands of the sample size and examination speed.

5. Conclusion

This paper has described a general method which is used to determine crystallographic orientation from the velocity of SAWs propagated in different directions, and that the SRAS

Journal of Physics: Conference Series **520** (2014) 012017 doi:10.1088/1742-6596/520/1/012017

(a) SRAS, pixel size $25\times7\mu$m.

(b) EBSD, pixel size $5\times5\mu$m.

Figure 3: Crystallographic orientation maps of a nickel sample by SRAS and EBSD techniques.

technique may be used to perform these measurements in a rapid and robust fashion. A SAW velocity model of crystals has been built according to Farnell's theory and this is matched with the measured velocity measurements to recover the orientation. In the present work the effect of generation and detection efficiency is considered in a rather *ad hoc* fashion and in future an explicit model of generation efficiency — for SAWs excited by a pulsed laser — will allow one to match the experimentally measured displacements without additional assumptions.

We have demonstrated that the SRAS technique is capable of being employed on multigrain samples. In order to establish the general applicability of the technique for cubic materials we have shown that the orientation of aluminium and Inconel 617 samples can be recovered with good precision. The SRAS technique also has been confirmed having a good agreement with EBSD by the experimental data of a nickel sample. The theoretical approach used in this paper can be applied to crystal types other than cubic and we propose to determine crystal type in materials with different phases such as titanium.

Acknowledgements

This work was supported by the EPSRC Platform Grant in Advanced Ultrasonics (EP/G061661/1). Thanks BJ Foss and BA Shollock for EBSD data, the E.ON New Build & Technology Limited for the Inconel 617 sample and the support from Rolls-Royce plc.

References

[1] Wickramasinghe H K and Hall M 1976 Phase imaging with the scanning acoustic microscope *Electronics Letters* **12** 637-638

[2] Li Wenqi 2012 Laser Ultrasonic Method for Determination of Crystallographic Orientation of Large Grain Metals by Spatially Resolved Acoustic Spectroscopy (SRAS) Ph.D thesis The University of Nottingham

[3] Li W, Sharples S D, Clark M and Somekh M G 2011 Frequency Spectrum Spatially Resolved Acoustic Spectroscopy for Microstructure Imaging *J. Phys.: Conf. Ser.* **278** 012-026

[4] Farnell G W 1970 *In Physical Acoustics — Principles and Methods* Mason W P and Thurston R N (Academic Press: New York and London) vol VI chapter 3 pp 109-166

[5] Li Wenqi, Sharples S D, Smith R J, Clark M and Somekh M G 2012 Determination of crystallographic orientation of large grain metals with surface acoustic waves *Journal of the Acoustical Society of America* vol 132 issue 2 pp 738-745

Journal of Physics: Conference Series **520** (2014) 012018 doi:10.1088/1742-6596/520/1/012018

Laser ultrasound and simulated time reversal on bulk waves for non destructive control

G Diot[1], H Walaszek[2], A Kouadri-David[3], S Guégan[3] and J Flifla[4]

[1]Institut Maupertuis, Contour St-Exupéry, Campus de Ker Lann, 35170 BRUZ, France
[2]CETIM de Senlis, 52 Avenue Félix - Louat - BP 80067 - 60304 SENLIS Cedex, France
[3]Université Européenne de Bretagne, INSA-LGCGM - EA 3913, 20, avenue des Buttes de Coesmes 35043 RENNES Cedex, France
[4]ECAM Rennes - Louis de Broglie, Campus de Ker Lann, 35170 BRUZ CS 29128, France

E-mail: gaeldiot@gmail.com

Abstract. Laser welding of aluminium generally creates embedded welding defects, such as porosities or cracks. Non Destructive Inspection (NDI) after processing may ensure an acceptable weld quality by defect detection. Nowadays, NDI techniques used to control the inside of a weld are mainly limited to X-Rays or ultrasonics. The current paper describes the use of a Laser Ultrasound (LU) technique to inspect porosities in 2 and 4-mm thick sheet lap welds. First experimentations resulted in the detection of 0.5-mm drilled holes in bulk aluminium sheets. The measurement of the depth of these defects is demonstrated too. Further experimentations shows the applicability of the LU technique to detect porosities in aluminium laser welds. However, as the interpretation of raw measures is limiting the detection capacity of this technique, we developed a signal processing using Time-Reversal capabilities to enhance detection capacities. Furthermore, the signal processing output is a geometrical image of the material's inner state, increasing the ease of interpretation. It is based on a mass-spring simulation which enables the back-propagation of the acquired ultrasound signal. The spring-mass simulation allows the natural generation of all the different sound waves and thus enables the back-propagation of a raw signal without any need of filtering or wave identification and extraction. Therefore the signal processing uses the information contained in the compression wave as well as in the shear wave.

1. Introduction

Assembling aluminium alloy can be achieved through various techniques. The absence of non-destructive control methods to quantify the assemblies' quality, can become a factor in choosing one or another technique. Laser welding in particular can generate different kinds of defects such as cracks or porosities, depending on which aluminium alloy is welded. The goal of our study is to explore different non-destructive controls so as to propose a novel method of inspection, which will be capable to detect on-line inner defects in an industrial context. For this purpose, we chose the ultrasound principle coupled with the laser ultrasound technique. This system has already shown capabilities of inner defects detection in surface treated aluminium alloys [1]. Detecting defects and assessing their sizes can also be improved by the use of signal processing techniques such as SAFT (Synthetic Aperture Focalisation Technique) [2]. To obtain a geometric image of the material's inner state through ultrasound measures, we chose to use Time-Reversal

Journal of Physics: Conference Series **520** (2014) 012018 doi:10.1088/1742-6596/520/1/012018

properties. These will determine the regions of interest within the measure. To perform this, the simulation tool seems to be quite adequate [3].

2. Simulation

The system we simulate is a possible description of a solid material expressed as a coupled mass-spring network [4]. We define $u_a = p_{a,b} - p_{0\,a,b}$ with p the position for a mass a and k the springs constant. We can write the following relationship in the case of mass-spring interaction:

$$\begin{aligned} F_a &= k\left[(u_{a+1} - u_a) - (u_a - u_{a-1})\right] \\ &= kl\left[\frac{(u_{a+1} - u_a)}{l} - \frac{(u_a - u_{a-1})}{l}\right] \end{aligned} \tag{1}$$

The notations can be found on figure 1.

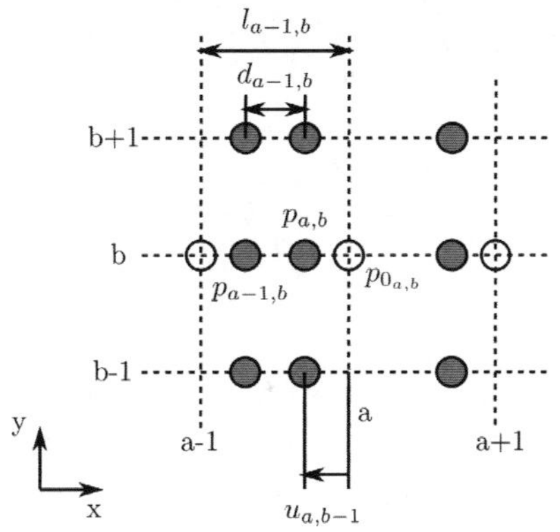

Figure 1: Initial state of the model's grid

When taking into account the limit on l and therefore consider a large amount of mass-spring systems connected to one another and then express the equation in the continuous domain and by defining $c^2 = (k/m) \times l^2$ as the wave's celerity to obtain the wave propagation equation [5]:

$$\frac{\partial^2 u}{\partial t^2} = c^2 \frac{\partial^2 u}{\partial x^2} \tag{2}$$

The relation between k, m, l, c and physical variables is:

$$c^2 = \frac{k}{m} l^2 = \frac{E}{\rho} \tag{3}$$

with E Young's modulus and ρ the volumetric mass of the considered material. It is so established that a mass-spring description can lead to the same well-known wave-propagation equation.

3. Experimental set-up

Ultrasonic measurements were performed with a pulsed laser to generate the soundwave and an interferometer to record the surface displacement induced by this soundwave. The pulsed laser interacts with the surface to produce ultrasonic surface and bulk waves. The interferometer measures the out-of-plane surface displacement. Each experiment was done in ablation mode so as to obtain a sufficient Signal-to-Noise Ratio (SNR) and an appropriate soundwave directivity.

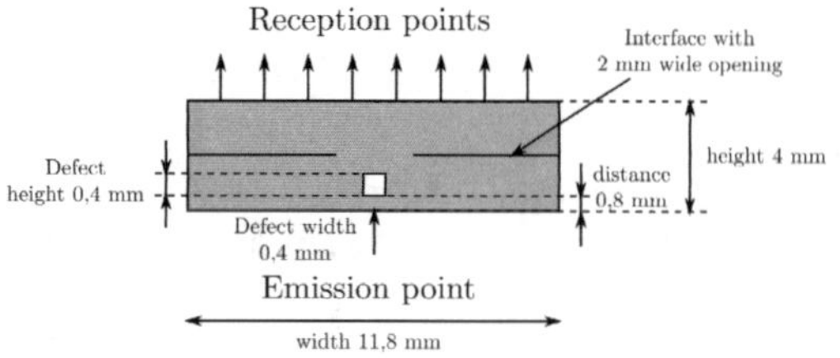

Figure 2: Configuration of the simulated plate with included defect

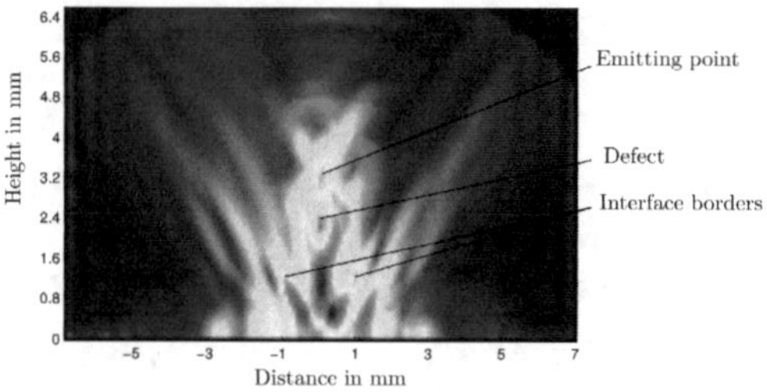

Figure 3: Extraction of the maximum energy for each point over the whole propagation duration

The impact can be approximated as a punctual point, thus defining the direction of the different waves. The pulsed laser is a Nd:YAG laser of wavelength $\lambda = 1064$ nm. The pulse length is 10 ns and therefore, the generated ultrasonic wave has a maximum frequency at -3 dB of 20 MHz. The interferometer uses a frequency doubled Nd:YAG laser at $\lambda = 532$ nm. Its frequency range of detection is: 20 kHz to 45 MHz. The measurement were performed with the excitation on the weld and the interferometer on the opposite surface. This configuration allow the detection of defect, but does not allow to precisely identify each porosity nor its size [6].

4. Time-Reversal

4.1. Time-Reversal Principle

The Time-Reversal principle has been established by Fink et al [7]. The ultrasound signal must be recorded by a receiver array. By sending the recorded ultrasonic signal backwards on the exact same position as the recording device's, the ultrasound signal will then use the path it took during the initial emission. Moreover, the backwards-emitted signals will also focus on the initial emitting point. In addition, the Huygens principle ensures that every interface can be considered as an emitter. The signal will not only naturally focus on the initial emitting point but also on every interfaces that stand on the wave's path. This property can then be used to detect defects.

4.2. Time-Reversal with numerical simulation input

We simulated the whole signal: both the initial emission and the time-reversed signal. The simulated configuration is two 2-mm thick, lap welded aluminium alloy plates with an included rectangular hole 0.4 mm wide and 0.8 mm deep (figure 2).

The displacement signal is only recorded out-of-plane on 120 points on the opposite side of the simulated plate, thus simulating a transmission configuration. Each recorded point is distant from the other by 0.1 mm. The recorded signal is then sent backwards on the same points, in a material with the same physical properties but without any prior geometrical information, neither on its size nor on the presence of a defect. The Time-Reversal is then performed. The signals are recorded on each point, for every iteration. The visualisation which is used consist in the extraction of the masses' kinetic and potential maximum energy ($\max(U = U_c + U_p)$), figure 3. The visualisation consists in storing the maximum value of the energy for each point over the whole back propagating duration. The results show the focus on the surface of the defect as well as on the emitting point. The maxima's extraction undoubtedly reveals the emitting point. Concerning the defect and the interfaces, the energy levels clearly shows the outline of the defect (figure 3) whereas for the interfaces, only the tips are being identified. Moreover one side does not show any potential identification ambiguity where the other can be misinterpreted because

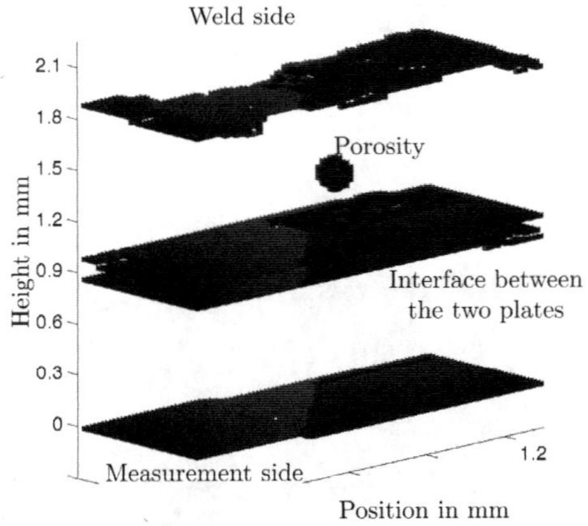

Figure 4: X-ray tomography of the tested lap-welded part

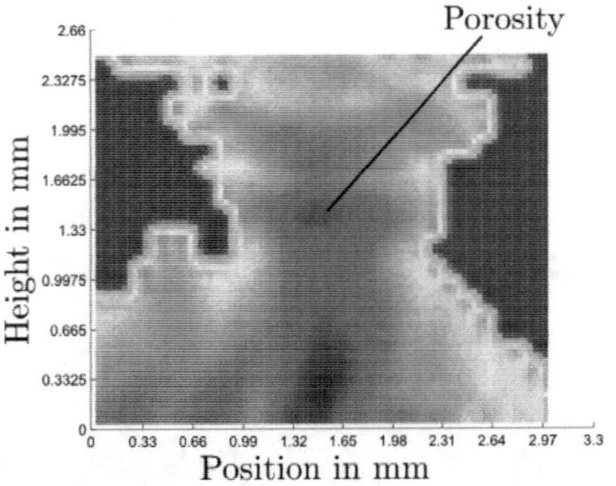

Figure 5: Maximum extraction after Time-Reversal with real data input

of an close energy peak. The Time-Reversal on simulated signals shows the effectiveness of this method even if some smearing signals can create false positive.

4.3. Time-Reversal with real data input

The next experiment uses real raw measurement data on lap-welded aluminium alloy plates. Only out-of-plane information was available and therefore sent backwards in the simulation. To compare the results, X-ray tomography has been used to characterise the tested part. The result is shown figure 4 and shows an included porosity in the upper plate.

The maximum extraction shows a maximum of energy where the porosity is. The porosity is smaller as the defect that has been simulated before therefore the porosity does not show the outline of the defect. Moreover some smearing is to be expected giving the high signal close to the emission points. This result is encouraging and more experiments should be conducted to confirm the method's effectiveness.

5. Conclusion

We hereby present a method that successfully reconstructed the internal status of lap-welded aluminium alloy plates. Furthermore, the prior knowledge about the inspected material to achieve such result is minimal. This method still needs improvement, the maximum energy extraction is sufficient in some cases but is not sensitive enough for the smallest defects. Nevertheless, simulated and experimental results show the effectiveness of this approach. Further development should include a computational optimisation to achieve a real-time analysis.

References

[1] Lévesque D, Mandache C, Dubourg L and Gougeon P 2008 Detectability of FSW defects with NDE methods *7th International Friction Stir Welding Symposium Proceedings: Papers on the Development and Application of Friction Stir Welding Presented at the Symposium Held in Japan*
[2] Lévesque D, Blouin A, Néron C and Monchalin J P 2002 *Ultrasonics* **40** 1057–1063
[3] Leutenegger T F 2002 *Detection of defects in cylindrical structures using a time reverse numerical simulation method* Ph.D. thesis Swiss Federal Institute of Technology Zurich
[4] Krautkrämer J and Krautkrämer H 1990 *Ultrasound testing of materials* 4th ed
[5] Kundu T 2003 *Ultrasonic nondestructive evaluation* 1 ed (CRC Press) ISBN 978-0849314629
[6] Diot G, Koudri-David A, Walaszek H, Guegan S and Flifla J 2013 *Journal of Nondestructive Evaluation* 1–8 ISSN 0195-9298 URL `http://dx.doi.org/10.1007/s10921-013-0189-5`
[7] Fink M 1992 *Ultrasonics, Ferroelectrics and Frequency Control, IEEE Transactions on* **39** 555–566

Journal of Physics: Conference Series **520** (2014) 012019 doi:10.1088/1742-6596/520/1/012019

Computer Simulation for Air-coupled Ultrasonic Testing

H Yamawaki

National Institute for Materials Science, 1-2-1, Sengen, Tsukuba, Ibaraki 305-0047 Japan

E-mail: yamawaki.hisashi@nims.go.jp

Abstract. Air-coupled ultrasound is used as non-contact ultrasonic testing method. For wider application of air-coupled ultrasonic technique, it is required to know situation of ultrasonic propagation between air and solid. Transmittance of the ultrasonic waves from air to solids is extremely small with 10^{-5}, however it was revealed that, by using computer simulation methods based on the two-stage elastic wave equation in which two independent variables of stress and particle velocity are used, visualization calculation of ultrasonic propagation between air and solid was possible. In this report, the calculation of air-coupled ultrasound using the new Improved-FDM for computer simulation of ultrasonic propagation in solids is shown. Waveforms obtained by 1-dimensional calculation are discussed for principle and performance of the calculation. Visualization of ultrasonic incidence to cylindrical steel pipe is demonstrated as an example to show availability for ultrasonic testing.

1. Introduction

Air-coupled ultrasonic technique[1,2] is utilized for non-destructive testing for advanced materials and structures. It requires no couplant other than air for ultrasonic transmission to the object of the testing. However, ultrasonic propagation between air and solid is extremely inefficient because of large difference in acoustic impedances of them. Moreover, their ultrasonic velocities differ extremely, therefore the propagation behaviour is complicated. Although the computer visualization is effective for analysis of ultrasonic behaviour, traditional method based on FDM(finite difference method) using displacement as independent variable could not treat directly ultrasonic propagation between air and solid. For the computer visualization, author was developed Improved-FDM which uses displacement as the variable[3], and recently the method was renewed in which particle velocity and stress are used as the variables[4]. Using former and new Improved-FDMs for visualization of ultrasonic propagation between air and steel, it was revealed that the new Improved-FDM could calculate ultrasound of air and steel in unified calculation model. In this report, principle of the calculation is explained and ultrasonic visualization for ultrasonic propagation between air and steel pipe is demonstrated.

2. Principle of FDM calculation for air-coupled ultrasound.

In comparison between ultrasonic physical quantities, ultrasonic transmittance between air and solid are attractive as shown in table 1 in which relations between air and steel are described. Between the group of stress and pressure and the group of particle velocity and displacement, they have complementary relationship. For example, in the case of propagation from air to steel, particle velocity decreases to 10^{-5}, whereas Stress increases to 2. This complementary characteristic plays important role in the calculation for ultrasonic propagation between materials with huge difference of acoustic impedance.

Journal of Physics: Conference Series **520** (2014) 012019 doi:10.1088/1742-6596/520/1/012019

Table 1. Ultrasonic transmittance between air and steel, in the case of normal incidence at room temperature. Z_1 and Z_2 is acoustic impedance of materials.

	Equations for transmittance from medium-1 to medium-2.	Transmittance from air to steel.	Transmittance from steel to air.
Stress, Pressure	$T_P = 2Z_2/(Z_2 + Z_1)$	1.999982	$1.76\text{x}10^{-5}$
Particle velocity, Displacement	$T_V = 2Z_1/(Z_2 + Z_1)$	$1.76\text{x}10^{-5}$	1.999982

For the computer visualization, volume of air and solid is discretized and elastic wave equation is transformed into finite difference formula. Figure 1 shows calculation model and elastic wave equations for Improved-FDMs. In former Improved-FDM, only displacement u is used as independent variable as shown in equation (3). In this case, subtraction in the formula can cause numerical error of "loss of trailing digits" or "loss of information". So, in the case of the propagation from air to solid, calculation can't reproduce extremely small displacements at solid surface, and in actual calculation, the transmittance was calculated to be 0. On the other hand, in new Improved-FDM, the two-stage elastic wave equation shown in equations (1) and (2) is used, and independent variables of particle velocity v and stress T are used. On each time-step of calculation, particle velocity and stress are calculated and updated by using each other. That is, each variable plays complementary role, and this process enables calculation of both transmissions between air and solid without loss of trailing digits. For example, in the case from air to solid, information of ultrasonic displacement is carried by mediation of the stress, without loss of information, as supposed on table 1.

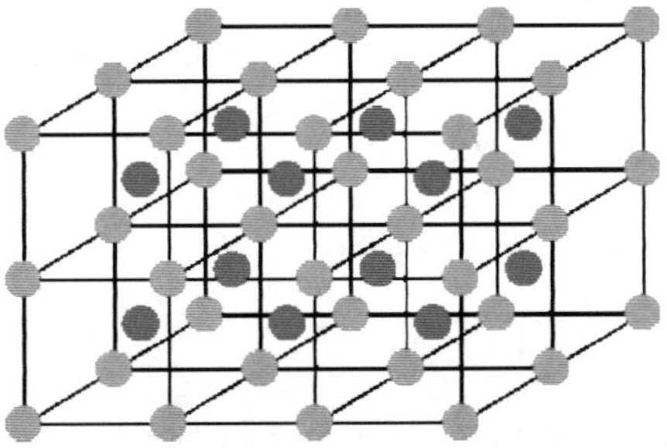

+

For new Improved-FDM:

$$\rho \frac{\partial v}{\partial t} = \frac{\partial T}{\partial x} \quad (1)$$

$$\frac{\partial T}{\partial t} = C \frac{\partial v}{\partial x} \quad (2)$$

For previous method:

$$\rho \frac{\partial^2 u}{\partial t^2} = C \frac{\partial^2 u}{\partial x^2} \quad (3)$$

Figure 1. Discretized solid model and wave equations used in Improved-FDMs[3][4]. Equations are of 1-dimension for simplicity. Corner points of unit cubes are calculation nodes for particle velocity v and displacement u, and the center points of unit cubes represent stress T, density ρ and elastic constant C.

Figure 2 shows waveforms obtained by 1-dimensional calculation of ultrasonic propagation from air to steel. In new Improved-FDM, particle velocity and stress are used and calculated. Displacement is obtained by time integration of particle velocity. The calculation was performed using the 3-dimensional program with cyclic boundary conditions. In the calculation, 2-cycles sinusoidal stress of 100kHz is loaded on one of particle velocity nodes of air. The excited displacement waveform is shown in the figure as the incident wave. Amplitude of waveform transmitted into steel is reduced to $1.75\text{x}10^{-5}$, which is good agreement with table 1. Surprisingly, degradation of transmitted waveform is not observed in spite of huge reduction of its amplitude, although increase of ringing trail is observed slightly. The ringing is the common phenomenon in FDMs and relating to propagation distance. The reflected waveform from steel surface shows that the steel acts as rigid surface where phase inversion and no amplitude reduction are occurred.

Journal of Physics: Conference Series **520** (2014) 012019 doi:10.1088/1742-6596/520/1/012019

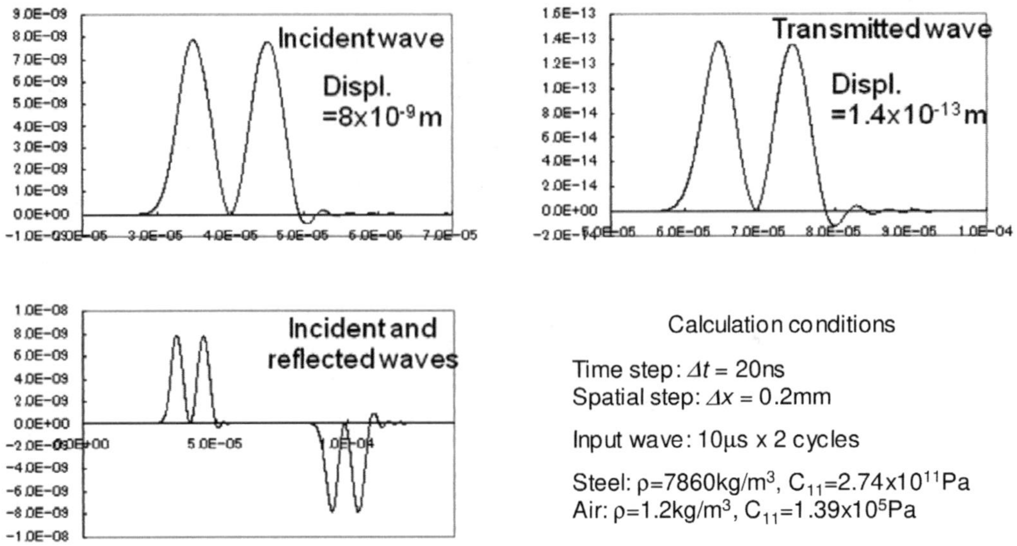

Figure 2. 1-dimensional calculation of ultrasonic propagation from air to steel. Vertical axis is displacement (m), and horizontal axis is time (s).

3. Visualizing calculation of ultrasonic propagation from air to steel pipe.

Important application of air-coupled ultrasonic is non-destructive testing for solid-fuel rocket[5]. Solid fuel with free from cracks and voids, and adhesion between the fuel and outer shell are important to assure normal combustion of the fuel. Requirement of no couplant and fire safety is remarkable advantage for the testing. However situation of ultrasonic propagation through rocket body was investigated only by the experiment. Here shows a simple calculation as an example for ultrasonic propagation into cylindrical body of the rocket.

Figure 3 shows 2-dimensional calculation for visualizing ultrasonic propagation from air to a cylindrical steel pipe. The calculation is performed using the 3-dimensional program with cyclic boundary condition. Two cycles of sinusoidal wave shown in figure 2 enters the steel pipe. The stiffness matrix of air used in the calculation is shown in equation (4). It is induced from Pascal's principle as well as for water. In the figure, displacement distribution of waves in the steel pipe can not observed on the images showing outer air (Air-o) and steel pipe because of amplitude reduction with 10^{-5}. On the other hand, on the images showing the steel pipe and inner air (Air-i), the distribution is observed clearly, and it is obvious that no reduction of displacement occurs while the wave propagates from steel to inner air. In the inner air, beam width of the longitudinal wave is observed wider in comparison with initial beam width. It is because that, in the case of the incidence from air to steel, mode conversion between longitudinal and shear wave occur with small angled incidence (~7deg.) and direct longitudinal wave and mode-converted waves are overlapped in the inner air.

$$C = \begin{bmatrix} 1.39e05 & 1.39e05 & 1.39e05 & 0 & 0 & 0 \\ 1.39e05 & 1.39e05 & 1.39e05 & 0 & 0 & 0 \\ 1.39e05 & 1.39e05 & 1.39e05 & 0 & 0 & 0 \\ 0 & 0 & 0 & 0 & 0 & 0 \\ 0 & 0 & 0 & 0 & 0 & 0 \\ 0 & 0 & 0 & 0 & 0 & 0 \end{bmatrix} \text{Pa} \qquad (4)$$

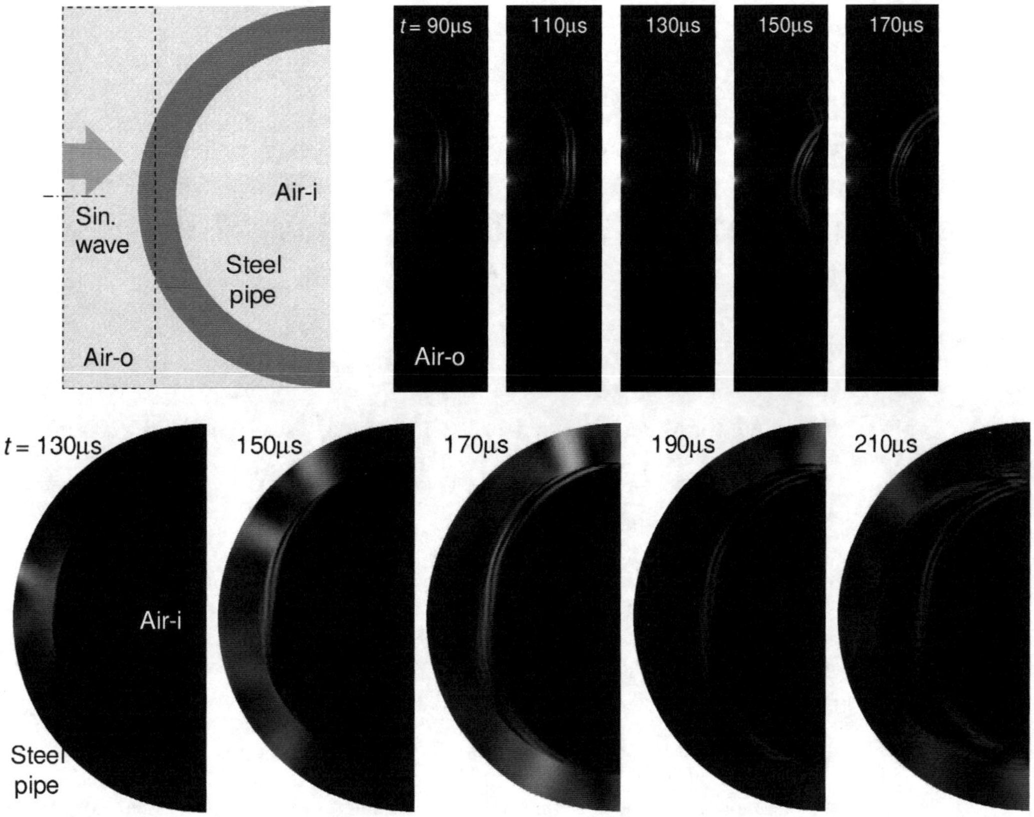

Figure 3. 2-D. calculation of angled-incidence of ultrasound into cylindrical steel pipe. Colors (red / green) in the images indicate directions of displacements (horizontal / vertical). Outside (upper images) and inside (lower images) of the pipe are visualized separately, because of huge difference of displacement amplitude. Steel pipe is 200mm outer diameter and 20mm thickness. Calculation condition is same as figure 2. Total area size is 700x1000 steps.

4. Conclusion

It was revealed that ultrasonic propagation between air and solid is calculated precisely using new Improved-FDM, because 2 independent variables of particle velocity and stress shows complementally relationship to prevent loss of trailing digit in the calculation. This feature is brought by the two-stage wave equation, therefore it is expected that the feature is also for FDTD[6] and EFIT[7].

References

[1] Schindel D W, Hutchins D A 1995 *Ultrasonics* **33** 11
[2] Schindel D W, Hutchins D A, Grandia W A 1996 *Ultrasonics* **34** 621
[3] Yamawaki H and Saito T 2000 *NDT&E International* **33** 489
[4] Yamawaki H 2010 *J. of the Japanese Society for Non-Destructive Inspection* **59** 624 (in Japanese)
[5] Minato M, Yamashita K, Siono H, Sato A, Sato E 2010 *J. of the Japanese Society for Non-Destructive Inspection* **59** 515 (in Japanese)
[6] Sato M 1995 *Jpn. J. Appl. Phys.* **34** 2808
[7] Fellinger P, Marklein R, Langenberg K J and Klaholz S 1995 *Wave Motion* **21** 47

1-Dimensional Analysis of Ultrasound at Closed Interface of Solid

H Yamawaki

National Institute for Materials Science, 1-2-1, Sengen, Tsukuba, Ibaraki 305-0047 Japan

E-mail: yamawaki.hisashi@nims.go.jp

Abstract. As a first step to investigate mechanism of nonlinear ultrasonic generation at closed cracks, computer simulation for ultrasonic propagation in 1-dimensional solid including closed interface was examined using Improved-FDM. Fundamental calculation model which described interaction between open / closure motion of the interface and ultrasonic stress was developed. In the model, compression stress is distributed over the entire solid, as motive force for closure of the interface. The interface is exhibited by the small region, and its open / closure are determined using calculated strain of the region. As a result, motion of the interface causing generation of saw-tooth like displacement waveform was observed. Amplitude modulation of displacement waveform was also observed, and it indicated possibility that small fluctuation of open / closure timing caused the modulation of the amplitude.

1. Introduction

Nonlinear ultrasonic generations observed when ultrasonic wave propagates through closed interface of solid[1,2] are utilized for detecting the closed cracks of materials[3,4]. In the generation, the transmitted / reflected waves are distorted, and include harmonic and sub-harmonic frequency components. It is assumed that the nonlinear waves are generated by crack open / closure motion during incidence of ultrasonic stress waves. Mechanism of the phenomena was already investigated by theory and computer simulation[5,6], but It seems not to be obvious yet. Noticeable point is that both of harmonic and sub-harmonic wave generations can be caused only by the crack open / closure behavior. If proper model of open / closure motion of interface could be suggested, it can be helpful for investigation of the mechanism. In this report, as a first step of investigation of ultrasonic behavior at closed cracks, computer simulation model for 1-dimensional ultrasonic propagation with closed interface was examined using new Improved-FDM[7].

2. Modelling of closed interface in 1-dimensional solid for FDM

Figure 1 shows basic concept of interaction between ultrasound and closed interface in 1-dimensional solid. In steady state, compression stress is distributed uniformly in the solid as motive force for closing the interface. Incidence of sinusoidal ultrasonic stress wave causes back and forth motion and elastic deformation of small region including the interface. When tensile stress of ultrasonic wave arrives at the region and it exceeds the compression stress, strain of the region becomes tensile and the interface begins opening. In the calculation of FDM, 1-dimensional solid is discretized as shown lower in figure 1. The discretized solid is constructed with unit cells which consist of calculation nodes of particle velocity and stress. One of the unit cells is assigned as the small region with the closed interface. Strain S_{uc} of the unit cell is used for judgment of open / closure of the interface.

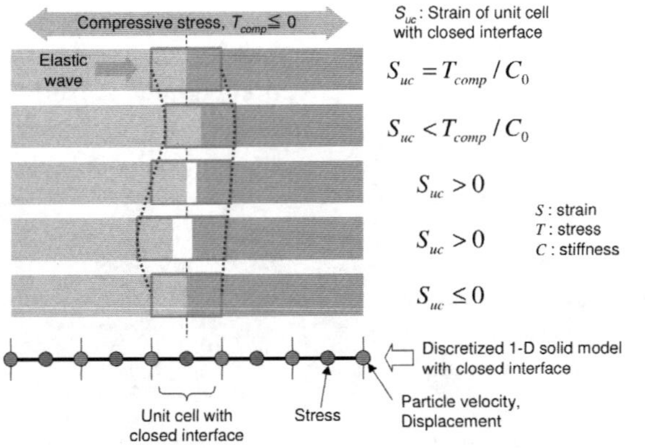

Figure 1. Concept of ultrasonic behaviour at the closed interface of 1-dimensional solid, and its calculation model for Improved-FDM. T_{comp} is initial compression stress distributed uniformly in solid. C_0 is elastic constant of solid.

For the calculation, finite difference equations based on 2-stage elastic wave equation are used. In the case of new Improved-FDM for 1-dimensional calculation, equations (1) and (2) are used to calculate time progress of each node. New values of stress T and particle velocity v are calculated from last values of each other for every time step. Suffixes (i), (iR) and (iL) indicate spatial position of nodes where iR: right position of i, and iL: left position of i. Δt, and Δx are time and spatial steps of FDM, C is elastic coefficient, and ρ is density. As shown in equations (3) and (4), displacement u is calculated by time integration of v, and strain S is calculated from u.

$$T_{new(i)} = (\Delta t/\Delta x)\cdot C_{(i)} \cdot (v_{old(iR)} - v_{old(iL)}) + T_{old(i)} \quad (1)$$

$$v_{new(i)} = 2/(\rho_{(iR)} + \rho_{(iL)})\cdot(\Delta t/\Delta x)\cdot(T_{new(iR)} - T_{new(iL)}) + v_{old(i)} \quad \text{: case of internal node} \quad (2)$$

$$v_{new(i)} = 1/\rho_{(iR)} \cdot (\Delta t/\Delta x)\cdot(T_{new(iR)}) + v_{old(i)} \quad \text{: case of left border} \quad (2a)$$

$$v_{new(i)} = 1/\rho_{(iL)} \cdot (\Delta t/\Delta x)\cdot(-T_{new(iL)}) + v_{old(i)} \quad \text{: case of right border} \quad (2b)$$

$$u_{new(i)} = \Delta t \cdot (v_{new(i)} + v_{old(i)})/2 + u_{old(i)} \quad (3)$$

$$S_{new(i)} = (u_{new(iR)} - u_{new(iL)})/\Delta x + S_{0(i)} \quad : S_{0(i)} = T_{comp}/C_{(i)} \quad (4)$$

Calculation flow of 1-dimensional new Improved-FDM with a closed interface is as follows. In initial condition, compression stress T_{comp} is set to every T of stress node in figure 1, as the force to close the interface, and every C is set to C_0. Body force of cosine wave ($=T_{amp}/2\cdot\cos(\omega t)$) is loaded at a particle velocity node in left area of the model, to generate sine wave of displacement. In every time-step of iterative calculation, when the strain of closed interface cell S_{uc} changes from negative value to positive value, the interface is judged to begin opening, and stress T and elastic coefficient C of the interface cell are set to 0. Then, when the strain changes to negative again, the interface is judged to begin closing, and the C is reset to C_0. This procedure is repeated during the calculation. Difference of calculations between closure and open states is on the calculation of 2 particle velocity nodes of the interface cell. In closure state, the calculations are performed using equation (2), while in open state, calculation is done using equations (2a) and (2b).

As shown in equations above, the new Improved-FDM uses independent variables of particle velocities and stresses, and this property enables simple introduction of the initial compression stress. Moreover, it is expectable that simple solid model of the Improved-FDM[7] makes future extension of the calculation into 3-dimension easy, although there is few difference between 1-dimensional calculation models of FDM-based calculation methods.

3. Calculation results of ultrasonic waveforms at closed interface.

Ultrasonic stress amplitude T_{amp} is important factor affecting to the nonlinear phenomena. Figure 2 shows displacement waves passing the interface and strain waveforms of the interface cell, relating to the T_{amp} in T_{comp}=-1Pa. Solid is steel of ρ=7860kg/m^3 and C_0=2.74x10^{11}Pa. In the case of T_{amp}=-1Pa, strain of the cell remains to be negative, therefore the interface remains closed and no deformation of displacement waveform occurs. In the case of T_{amp}=-10Pa and T_{amp}=-100Pa, the interface opens periodically, and saw-tooth displacement waveforms are generated. Attention should be paid to that amplitude of the saw-tooth waveforms is not proportional to T_{amp}.

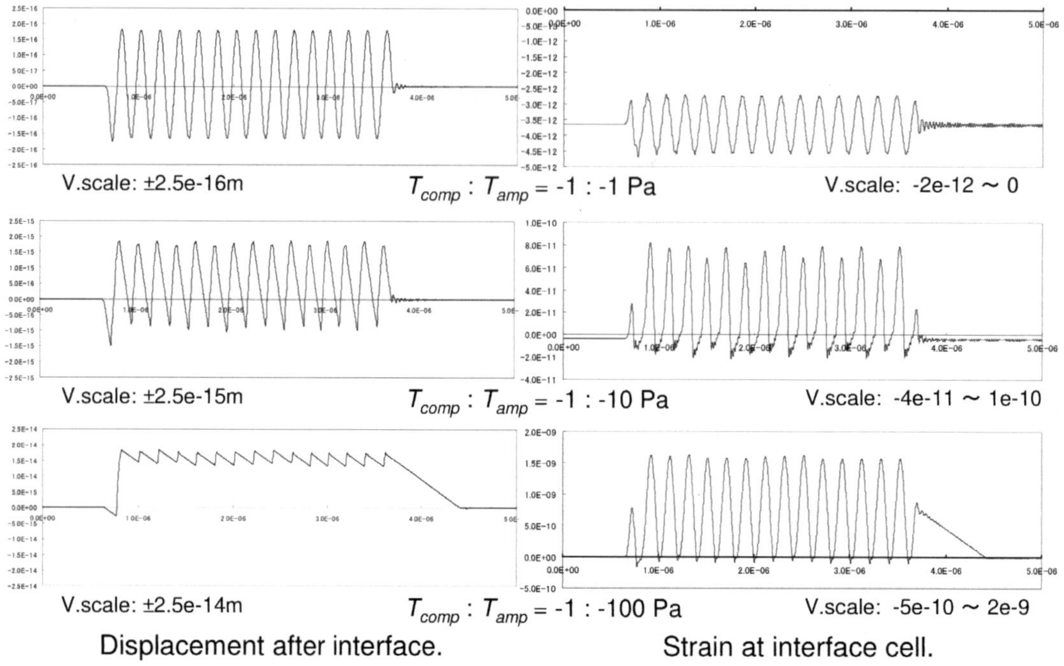

Figure 2. Ultrasonic waveforms of the interface cell. Using 15 cycles cosine wave with 5MHz of body force, sine wave of displacement is excited. Force amplitudes T_{amp} are -1Pa, -10Pa and -100Pa, and T_{comp}=-1Pa, Δt=2.5ns and Δx=40μs. Horizontal axis is time (s).

Figure 3 shows detailed waveforms of T_{amp}=-10Pa in figure 2. Waveforms of 2 displacement nodes of the interface cell (D_L, D_R) and waveforms corresponding to strain of the cell (=(D_R-D_L)/ Δx) are shown. The displacement waveform in figure 2 corresponds to D_R. It is obvious that, during the interface opens (D_R-D_L>0), D_R shows constant velocity of $V=T_{comp}V_L/C_0$ where V_L is longitudinal wave velocity, and composes the saw-tooth shape. Another noticeable point is slight amplitude modulation for T_{amp}=-10Pa. It's influence appears remarkably as in the strain waveform. By additional calculations, it was revealed that aspect of the modulation varies with calculation conditions and it is chaotic. From figure 3, it is assumed that fluctuation of open / closure timing of the interface is induced by excessive ringing waveform caused generally by frequency dispersion characteristic of FDM, and the fluctuation causes the modulation.

Figure 4 shows the reflected displacement wave at the cell. Fine calculation steps of Δt=0.5ns and Δx=10μs are used to obtain accurate waveforms. On the reflection, incident and reflected waves are overlapped on the node of D_L. To observe fine waveform, reflected waveform is obtained by subtraction between the overlapped and incident waves. As for the reflected displacement wave, it has good agreement with theoretically obtained waveforms[5], and amplitude modulation is also observed.

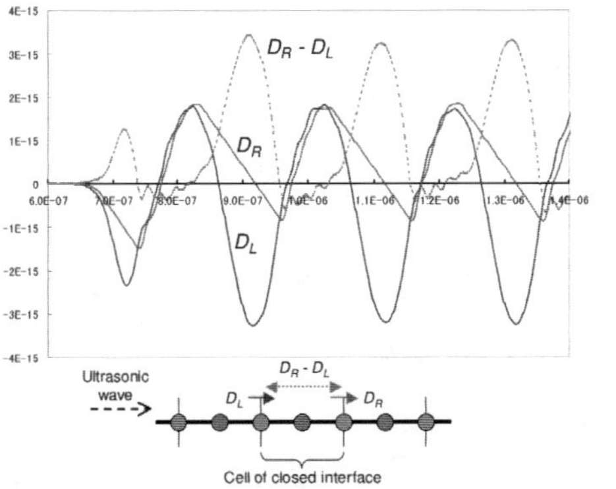

Figure 3. Waveforms of 2 displacement nodes (D_R, D_L) and strain (D_R-D_L) of the cell.

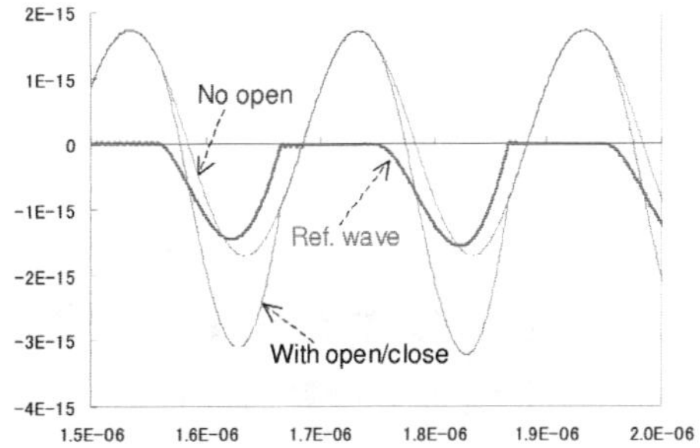

Figure 4. Reflected waveform obtained by subtraction between D_L and incident wave. Calculation steps are Δt=0.5ns and Δx=10μs. Other conditions are in accord with figure 3.

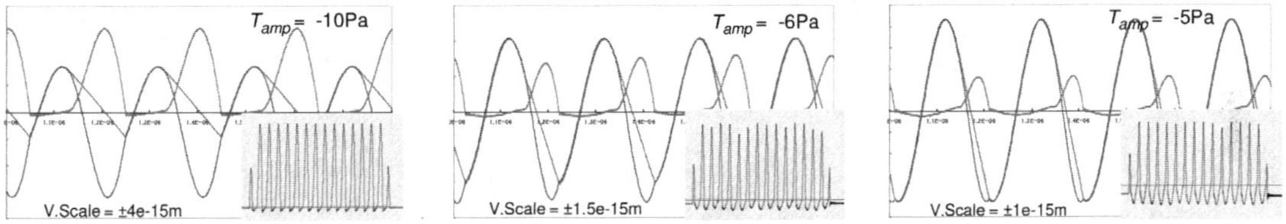

Figure 5. Waveforms in the same way with figure 3. Calculation steps are refined by Δt=0.5ns and Δx=10μs. Whole of strain waveform is appended to observe degree of the amplitude modulation. Force amplitudes T_{amp} are -10Pa, -6Pa and -5Pa.

Figure 5 shows accurate waveforms of figure 3 and the variations. Using fine calculation steps of Δt=0.5ns and Δx=10μs, the excessive ringing wave is reduced, and therefore the amplitude modulation is suppressed in the case of T_{amp}=-10Pa. However in the waves of T_{amp}=-6Pa and -5Pa, irregularities of displacement waveform are still detected by the observation of strain waveforms. It shows that even by the quite small fluctuation of the waveform, the amplitude modulation can occur unpredictably.

4. Conclusion

Computer simulation for ultrasonic propagation in 1-dimensional solid including the closed interface was examined using new improved-FDM. In the calculation, saw-tooth like displacement waveforms were observed and its amplitude modulations were also observed. It is sure that the saw-tooth waveform generation is the source of harmonic generation, and the modulation indicates possibility that small fluctuation of open / closure timing of the interface causes the sub-harmonic generation. In the calculation, source of the fluctuation was numerical error induced by characteristic of FDM. Further investigation will be required to identify the source of fluctuation in the real situation.

References

[1] Buck O, Morris W L, Richardson J M 1978 *Appl. Phys. Lett.* **33** 371
[2] Solodov I Y, Vu C A 1993 *Acoust. Phys.* **39** 476
[3] Kawashima K, Murase M, Yamada R, Matsushima M, Uematsu M, Fujita F 2006 *Ultrasonics* **44** e1329
[4] Ohara Y, Mihara T, Sasaki R, Ogata T, Yamamoto S, Kishimoto Y, Yamanaka K 2007 *Appl. Phys. Lett.* **90** 011902
[5] Richardson J M 1979 *Int. J. Engng. Sci.* **77** 73
[6] Yamanaka K, Ohara Y, Oguma M, Shintaku Y 2011 *Appl. Phys. Express* **4** 076601
[7] Yamawaki H 2010 *J. of the Japanese Society for Non-Destructive Inspection* **59** 624 (in Japanese)

Journal of Physics: Conference Series **520** (2014) 012021 doi:10.1088/1742-6596/520/1/012021

Enhanced Higher Harmonic Imaging of Heterogeneities and Local Plastic Deformation in Steel Plates

K Kawashima[1,2], R Imanishi[1], T Aida[3] and Y Zhou[3]

[1] Ultrasonic Materials Diagnosis Lab., Hazamaco-4, Nagoya, 466-0062, Japan
[2] Nagoya Industrial Science Research Institute, Yotsuya 1-13, Nagoya, 466-0819, Japan
[3] Insight K. K., Hyakunincho 2-6-7, Tokyo, 167-0073, Japan

E-mail: kkawas@nl-ultrasonics.co.jp

Abstract. Conventional ultrasonic imaging based on the difference in acoustic impedance fails to detect and visualize small heterogeneities and local plastic deformation in metals. Nonlinear ultrasonic imaging technique visualizes higher harmonic amplitudes which are generated at the heterogeneities by finite amplitude sinusoidal burst waves, therefore, it can be applied for detecting small non-metallic inclusions, local plastic deformation and micro cracks. By transmitting 35 MHz sine burst waves and receiving harmonics of 105 MHz in the maximum, non-metallic inclusions in stainless steel of some ten μm in size and crack tip plastic zone of 2 mm in diameter are visualized.

1. Introduction

Nonlinear acoustic or ultrasonic methods which make use of the waveform distortion from the incident sinusoidal wave are useful for detecting kissing bonds with small difference in acoustic impedance and minute defects within bond layers between dissimilar materials with the large impedance difference. Among them, a higher harmonic method quantitatively evaluates the waveform distortion by higher harmonic amplitudes.

The first author and et al. have developed an immersion higher harmonic imaging system and visualized closed cracks [1], micro voids in diffusion bonded interface [2], non-metallic inclusions [3], and local plastic deformation [4]. In these images, the maximum frequency of incident burst wave was 20 MHz and the third harmonic (60 MHz) amplitude was mapped. The incidence of higher frequency burst wave can enhance the higher harmonic amplitudes due to stronger cyclic stress excited within samples and result in higher spatial and temporal resolution by the short wavelength.

In the present work, we demonstrate higher harmonic images of non-metallic inclusions in stainless steel and local plastic deformation around a hole and crack tip by transmitting 35 MHz sine burst wave and receiving 80- 100MHz harmonic amplitudes.

2. Immersion higher harmonic imaging system

We use the experimental setup shown in Figure 1. The key components are a 100 MHz sine burst wave pulser (RITEC RAM-5000), an amplifier (RITEC RPR-4000), multi-selectable high-pass filters, a scanner and imaging software (InsightScan and InsightAnalysis). Samples were placed horizontally within the water bath, and the water path length is kept constant during scanning, therefore, higher harmonic amplitude exited in water is unchanged during scanning.

Journal of Physics: Conference Series **520** (2014) 012021 doi:10.1088/1742-6596/520/1/012021

We can use either the backscattered mode or the local resonance mode shown in Figure 2 for imaging defects or heterogeneities. By the back scattered mode, we have high spatial and temporal resolutions. On the other hand, we have high sensitivity for harmonic generation by the local resonance mode.

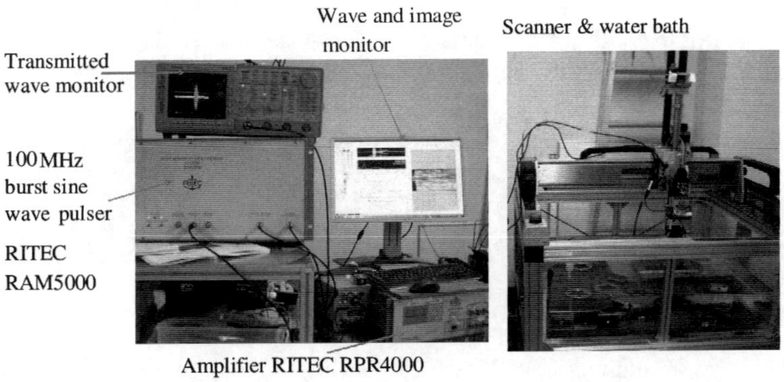

Figure 1. Immersion higher harmonic imaging system.

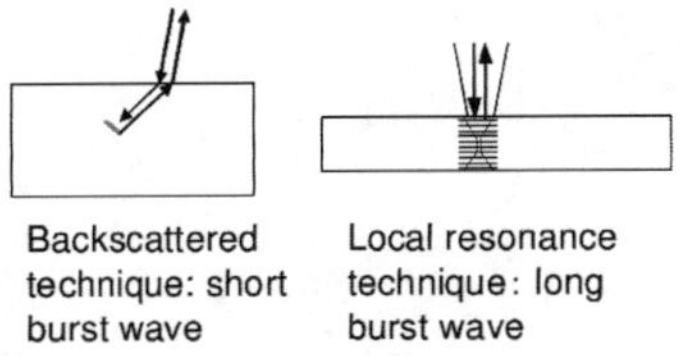

Figure 2. Backscattered and local resonance modes.

3. Higher harmonic images of non-metallic inclusions in stainless steel

Micro fatigue cracks often initiate at subsurface non-metallic inclusions in thin metal parts subjected to cyclic loading. Therefore, a reliable non-destructive technique is wanted for detecting such subsurface inclusions. The harmonic and conventional UT images of non-metallic inclusions in a SUS 304 plate are shown in Figure 3. The harmonic image was obtained with a focus transducer of 30 MHz in frequency and 25 mm in focal length by transmitting 35 MHz and 4 cycle sine waves and receiving mainly 105 MHz harmonic component with a high-pass filter of 80 MHz. For the conventional UT image, we used a spike wave pulser (Panametrics 5900PR). The received wave amplitude at every point is transformed in colour according to the colour scale shown in the middle.

In the harmonic B- and D-scan images, we clearly identify each inclusion shown in the square, while the identification of the inclusions shown in the circle is difficult in the conventional UT B- and D-scan images due to strong grain scattering noise. By examining SEM images and EPMA map, which are not shown, of the cross section, we found the typical inclusions were alumina and manganese sulphide elongated in the rolling direction.

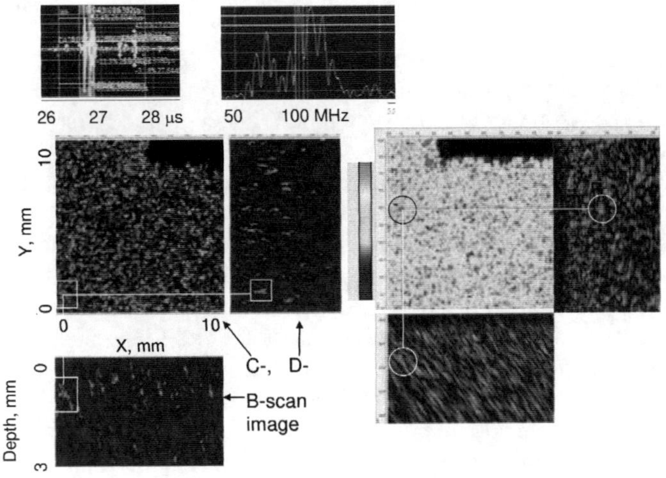

(a) Harmonic image (b) conventional UT image

Figure 3. Images of non-metallic inclusions in SUS 304 plate.

Journal of Physics: Conference Series **520** (2014) 012021 doi:10.1088/1742-6596/520/1/012021

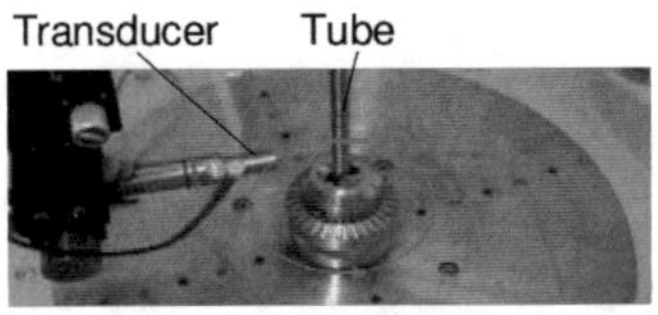

Figure 4. Experimental setup for ODS fuel cladding tube.

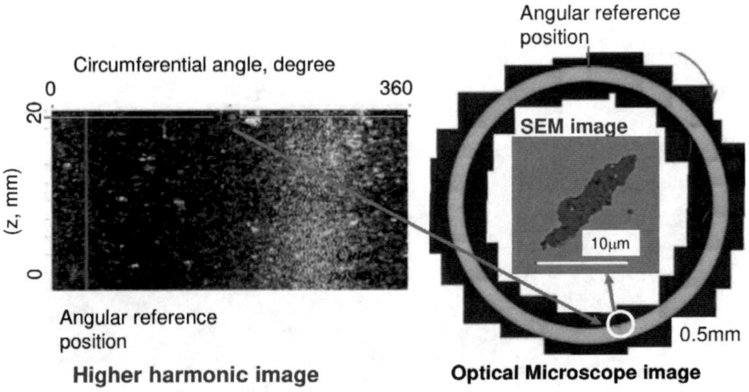

Figure 5. Harmonic, optical microscope and SEM images of an inclusion in ODS fuel cladding tube [5].

The sizes of inclusions imaged by higher harmonics are over 50 μm.

Another higher harmonic image of non-metallic inclusion in oxide dispersion strengthened (ODS) fuel cladding tubes for fast breeder reactors is shown in Figure 4. The material is a 9 Cr ferrite stainless steel strengthened by yttria (Y_2O_3) and is made at Japan Atomic Energy Agency. The harmonic image was taken by using the experimental setup shown in Figure 4. The tube of 6.5mm in diameter and 0.5mm in thickness was rotated circumferentially. After every rotation, the transducer was lifted axially by 0.1 mm. By this circumferential and axial scanning, circumferentially developed image of heterogeneities was constructed and it is shown in the left side of Fig. 5. Incident frequency is 20 MHz and 40 MHz high-pass filter is used to extract the 3rd harmonic amplitude.

Among several bright spots in the harmonic image, the spot shown in red circle was selected for detailed examination. After cutting the tube along the white horizontal line, polishing and observing by an optical microscope, we found an object shown in white circle and examined it by SEM. As shown in the central area, the object is found to be a TiO_2inclusion of 12×3μm in size.

At other 7 points where the higher harmonic amplitude is much higher than the matrix, heterogeneities such as oxide, aggregate of micro voids and Cr-rich segregation are confirmed by SEM and EPMA. Those sizes are smaller than 30 μm, therefore they are allowed for service. By the way, this cladding tube is fabricated by the powder metallurgy technique in extraordinary clean atmosphere, therefore, non-metallic inclusions are far less than those in common steel.

4. Higher harmonic imaging of local plastic deformation

Local plastic deformation around a circular hole in SUS 304 plate which was subjected to 30% plastic strain at the maximum is imaged by the local resonance mode. Incident sine burst wave is 22.35 MHz in resonance frequency and 70 cycles. With a high pass filter of 80 MHz, the 4th harmonic amplitude is mapped. As shown in Figure 6, at the just below the hole around the minimum cross section, the local plastic zone is clearly shown in the C-scan image shown in the top. This image is constructed by the wave amplitude in the white frame at the bottom. Temporal variation of the harmonic amplitude along the red horizontal line is shown in the middle, from which we notice that the delayed wave amplitude is large in the severely deformed plastic zone.

The harmonic image of plastic zone in front of fatigue crack tip and Luders band in a SPC plate of 1mm in thickness is shown in Figure 7. The incident sine burst wave is 21.26 MHz in resonance frequency and 80 cycles. With a high pass filter of 80 MHz, the 4th harmonic amplitude is extracted and it is shown in the middle. We can identify the crack tip plastic zone of approximate 3 mm in diameter and Luders bands. We notice that the delayed wave amplitude around 38.3 μs is large and the peak spectrum is about 85 MHz (4th harmonic) in the left figure.

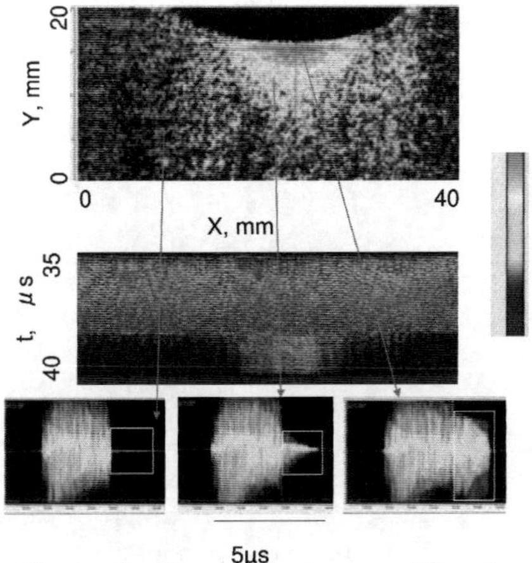

Figure 6. Harmonic image of local plastic zone in SUS 304 plate.

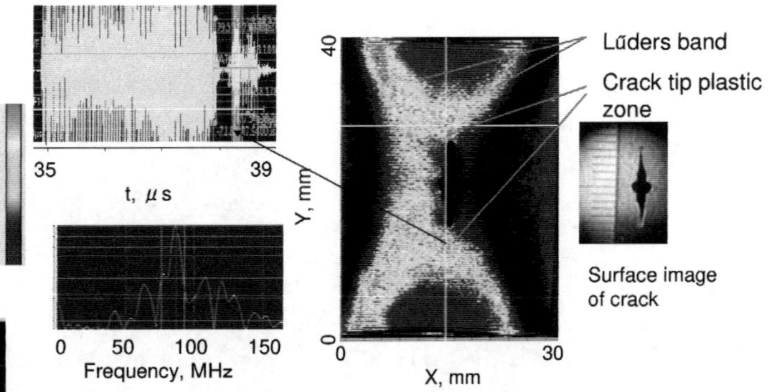

Figure 7. Harmonic image of crack tip plastic zone and Luders bands in thin steel plate.

5. Concluding remarks

With the immersion higher harmonic imaging system, we have non-destructively imaged non-metallic inclusion of 12×3μm in size within ODS fuel cladding tube by transmitting 20 MHz sine burst wave and receiving 60 MHz component. In addition, crack tip plastic zone of approximate 3mm in diameter is imaged by transmitting sine burst wave of 21.26MHz in resonant frequency and receiving the 4th harmonic amplitude. These have never been visualized by the conventional ultrasonic imaging.

The harmonic imaging of small non-metallic inclusions and small crack tip plastic zone could lead to evaluate nondestructively material degradation and damage evolution such as fatigue crack growth in the early stage of fatigue life. As pointed by Mayendorf et al. [6], nonlinear acoustic or ultrasonic technique is best fitted for non-destructive evaluation of the initial and middle stage of the crack growth.

Thus, the immersion higher harmonic imaging technique would be indispensable for non-destructive material characterization and evaluation in the future.

Acknowledgements

This work is partially supported by KAKENHI grant-in-aids for Scientific Research (C) (24560119).

References

[1] Kawashima K, Murase M, Yamada R, Matsushima M, Uematstu M and Fujita F 2006 *Ultrasonics* **44** e1329.
[2] Yamada R and Kawashima K 2006 *Res. Nondestr. Eval.* **17** 223.
[3] Kawashima K Ito T and Nagata Y 2010 *Jpn. J. Appl. Physics* **49** 7CH11.
[4] Kawashima K 2011 *Jpn. J. Appl. Phys*, **50** 07HC14.
[5] YanoY, Kawashima K, Tanno T, Ohtsuka S and Kaito T Proc. 2013 AESJ Autumn meeting 530
[6] Mayendorf N, Nagy P and Rokhlin S Eds. 2004 *Nondestructive Materials Characterization with applications to Aerospace Materials* (Berlin Springer) p 20.

Atypical Applications for Gas-coupled Laser Acoustic Detection

JN Caron,[1,2] **P Kunapareddy**[1]
[1]Research Support Instruments, 4325-B Forbes Boulevard, Lanham, MD, 20706, USA,
[2]Quarktet, Silver Spring, MD 20901, USA

E-mail: Caron@RSImd.com

Abstract. Gas-coupled laser acoustic detection (GCLAD) was primarily developed to sense laser-generated ultrasound in composite materials. In a typical setup, a laser beam is directed parallel to the material surface. Radiated ultrasound waves deflect or displace the probe beam resulting from changes in the air's index of refraction. A position-sensitive photodetector senses the beam movement, and produces a signal proportional to the ultrasound wave. In this paper, we discuss three applications of GCLAD that take advantage of the unique detection characteristics. Directivity patterns of ultrasound amplitude in water demonstrate the use of GCLAD as a directional hydrophone. We also demonstrate the sensing of waveforms from a gelatin. The gelatin mimics ultrasound propagation through skin tissues. Lastly, we show how GCLAD can be used as a line receiver for continuous laser generation of ultrasound. CLGU may enable ultrasound scanning at rates that are orders of magnitude faster than current methods.

1. Introduction

Gas-coupled laser acoustic detection (GCLAD) was originally developed to sense laser-generated ultrasound in composite materials. [1] In a typical setup, a laser beam is directed parallel to the material surface. As ultrasound travels through the material, a portion of the waveforms traverses the solid/air interface and radiates into the surrounding air. The airborne ultrasound waves create a deflection or displacement of the probe beam [2] resulting from changes in the index of refraction. A position-sensitive photodetector senses the beam movement, producing a signal proportional to the ultrasound wave. Since the technique is independent of the optical surface properties, it has been developed as an alternative to interferometric-based detection systems for inspecting materials with optically rough or nonreflective surfaces.

GCLAD has several unique properties that may prove useful in other acoustic sensing situations. The method requires very few components, has modest laser power and stability requirements, and has a broadband frequency response. The technique can be easily modified for different situations. Here we report on progress on three atypical applications of GCLAD:

- Using optical beam deflection or displacement to sense ultrasound in water.
- Sensing ultrasound transmitted through gelatins.
- Using GCLAD as a line receiver for detecting waveforms generated by Continuous Laser Generation of Ultrasound (CLGU). [3, 4, 5]

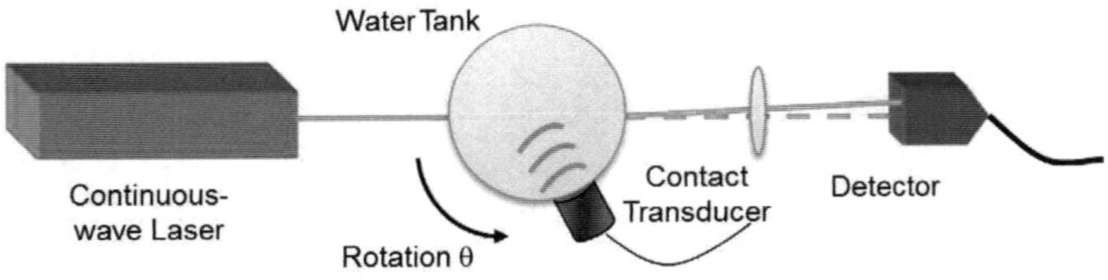

Figure 1. Experimental arrangement for the measurement of ultrasound directivity dependence in water.

2. Underwater Acoustic Detection

In recent research, [6, 7] laser-generated ultrasound has been created in water tanks and open water by creating an underwater plasma. When a laser pulse of sufficient energy enters the water, the pulse of light self-focuses, igniting a plasma. The expansion of the water as a result of the plasma produces the ultrasound. Acoustic waves with frequencies around 30 kHz were recorded in open water as far as 300 meters away with a conventional hydrophone.

In place of a conventional hydrophone, a laser-based detection system could be used to create a complete laser-based ultrasound system. A GCLAD-like detection method could have advantages over hydrophone detection, such as directivity, frequency range, and cost. In this paper, we report measurements of the directivity response of GCLAD for underwater ultrasound. If we can show that the directional dependence is strong in comparison to current directional hydrophones, this technique could be used as a highly sensitive directional hydrophone.

Figure 1 shows the experimental arrangement. A 1 MHz contact transducer was put in contact with a cylindrical tank filled with tap water. The polyethylene terephthalate (PET) tank is transparent to visible light and has a diameter of 22 centimeters. The tank was positioned on a rotation stage, allowing the transducer to be rotated along with the tank. A laser beam was directed through the tank such that it would encounter the ultrasound in the center of the

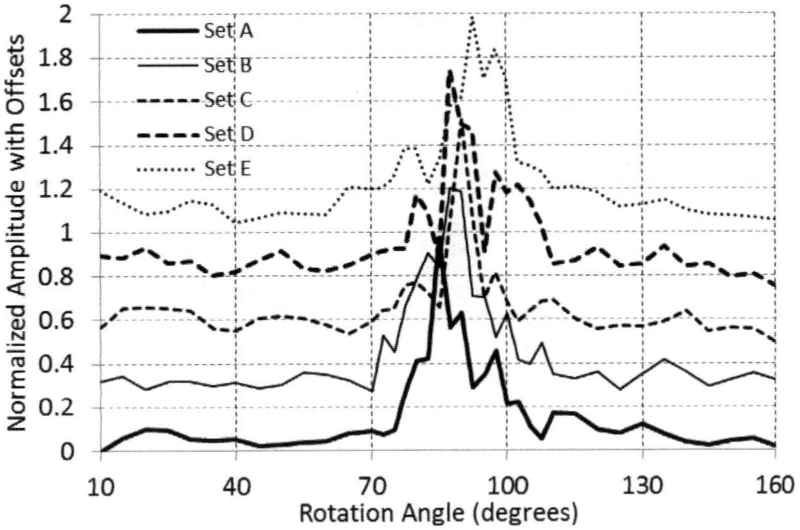

Figure 2. Directivity patterns for several experimental runs. Offsets were added to the data to better show the shifts of the central peak.

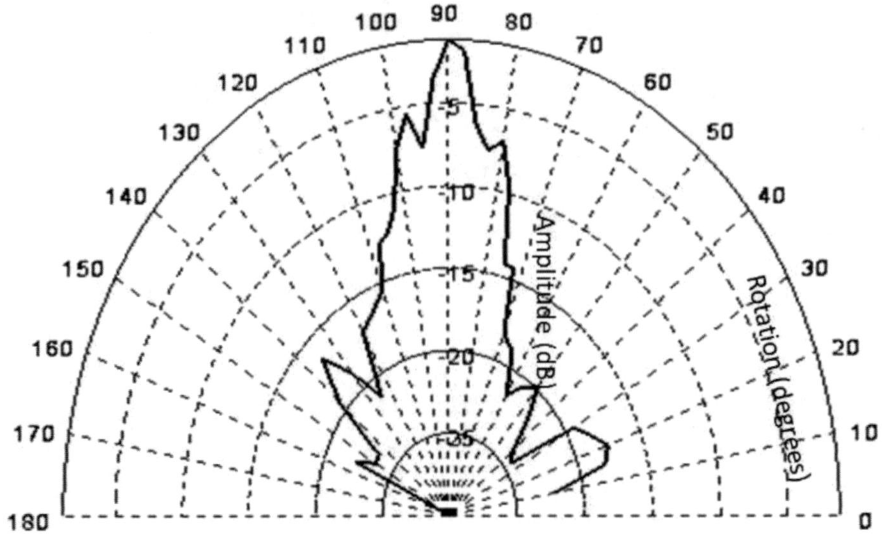

Figure 3. Polar plot for the average of four directivity tests.

tank and received through a lens by the position-sensitive detector. The placement of the lens, which is critical to achieve optimum sensitivity, was calculated using the method described in Caron. [8]

To form a directivity pattern, waveforms are captured for each angle. The peak-to-peak amplitude is taken from the first arrival of the longitudinal wave and plotted as a function of angle. Figure 2 shows the results of several experimental runs. The waveform is highly directional, with the majority of acoustic energy occurring within 20 degrees of the normal direction. A common figure of merit is to measure the width of the directivity pattern at half the maximum power. For the displayed runs, the range of "degrees at half power points" varies from $\pm$ 2.9 to 7.3 degrees and average to $\pm$ 5.4 degrees. For comparison, the Model DH-4 Directional Hydrophone, produced by Sonotronics, [9] specifies $\pm$ 6 degrees at half power points, however there was no information provided about the ultrasound frequency.

The shifts in the central peak shown in Figure 2 are most likely produced by day-to-day shifts in the experimental apparatus that occur when, for example, new couplant was placed on the transducer. To produce an average, the curves were shifted by degrees according to the location of their centroid. The average directivity pattern is shown in Figure 3 as a polar plot. The averaged curve shows a value of $\pm$ 5.9 degrees at half power points.

3. Laser-based Detection of Ultrasound in Gelatins

When studying anomalies in skin tissues, a thick gelatin can be used in place of the tissue for laboratory research. [10] These gelatins are created inexpensively and embedded with foreign objects to mimic different skin tissues and conditions. Various techniques have been used to generate and detect ultrasound in gelatinous materials. Generation is typically accomplished using a contact transducer or an array of transducers, but there has been research with laser generation as well. [11, 12] To sense the ultrasound, contact transducers are commonly used, but research has been performed with noncontact methods, such as noncontact piezoelectric transducers, [13] and interferometers. [14, 15]. A noncontact method may be preferable to contact transducers in situations where couplant cannot be applied, as is the case with severe burns.

Compared to the methods listed above, GCLAD has a much broader frequency response than standard piezoelectric transducers, and is less complicated to implement than an interferometer-

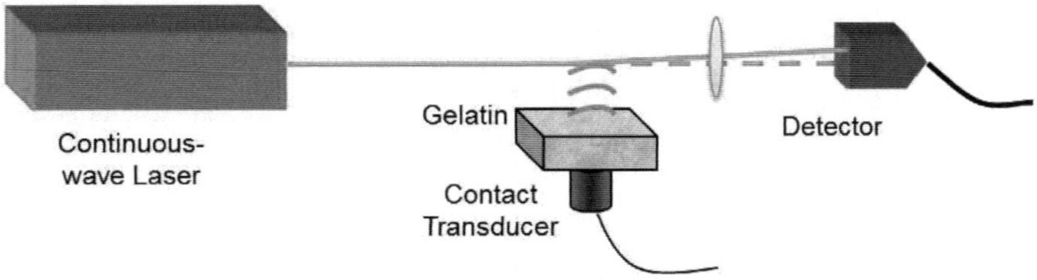

Figure 4. Laser Ultrasound Setup for sensing ultrasound in gelatins.

based system. As an initial test (based on a suggestion from a colleague [16]), a Jell-O^{TM} brand gelatin snack pack was inserted into the GCLAD system. A 1 MHz contact transducer was used to launch the ultrasound into the approximately 5 cm think sample, as depicted in Figure 4. The probe beam intercepts the waveform after it has radiated about an centimeter into the air on the opposite side of the transducer.

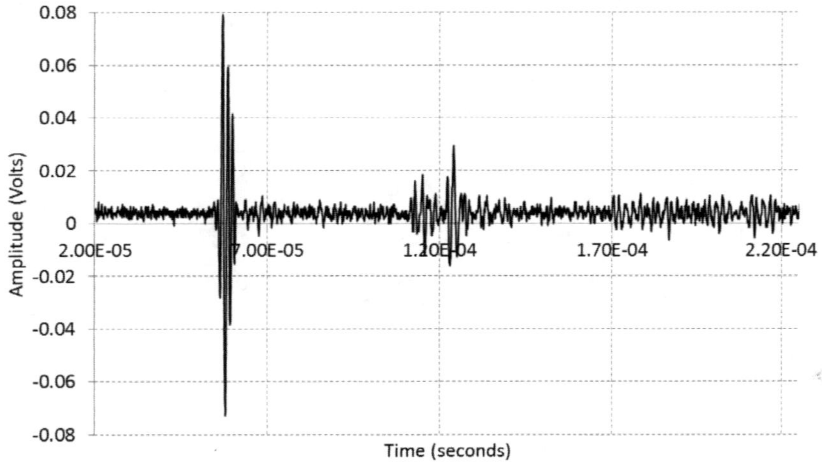

Figure 5. Longitudinal wave and the first echo in a 5 cm thick cube of Jell-O^{TM} brand gelatin, as recorded using GCLAD.

An example of a received waveform, averaged over 16 shots is shown in Figure 5. The first arrival is clearly resolved at 55.9 microseconds and an echo can be seen at 123 microseconds. The signal-to-noise ratio, found by dividing the peak-to-peak amplitude by the standard deviation, is 102.2 with averaging or 25.6 for a single shot.

A more suitable gelatin was created using a recipe by Bude and Adler [17]. According to the authors, this type of gelatin is useful for training for in vivo biopsy procedures. The gelatin had a thickness of 3 cm, and was much firmer than our original sample. The gelatin was inserted into the setup with a 4 cm air gap between the sample and the probe beam. Figure 6 (left) shows a received waveform, averaged over 16 shots, that was generated by putting a 500 kHz pulse into a 1 MHz transducer. Based on the arrival time of 13.2 μs and the speed of sound in air, the speed of sound in the material is 1652 m/s. A Fourier frequency transform of the wave reveals a central peak at 390 kHz (right).

Journal of Physics: Conference Series **520** (2014) 012022 doi:10.1088/1742-6596/520/1/012022

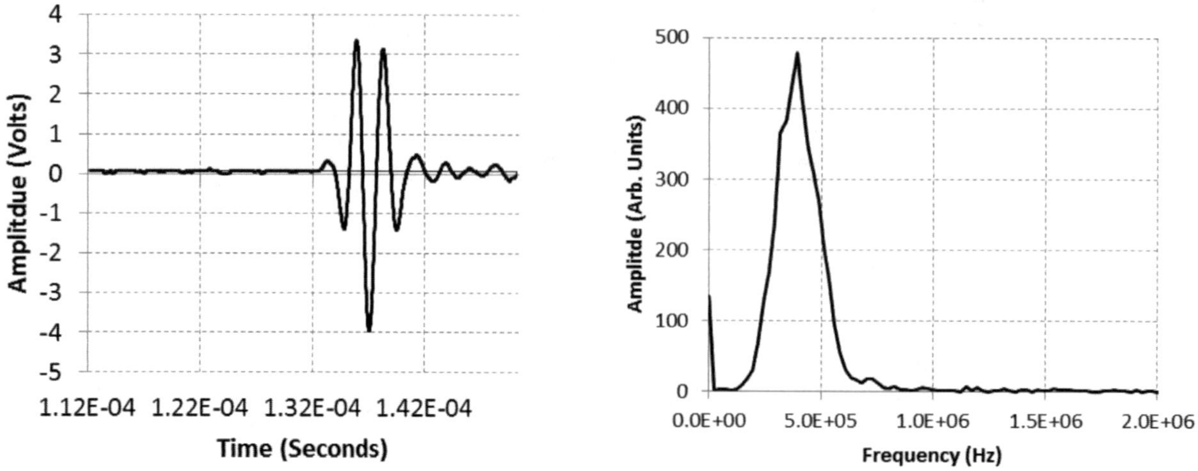

Figure 6. (Left) A longitudinal wave sensed by GCLAD after transmission through a 3 cm thick gelatin and a 4 cm air gap. (Right) An FFT of the waveform reveals a central peak at 390 kHz.

4. Line Receiver detection for CLGU

Continuous Laser Generation of Ultrasound (CLGU) has the potential to greatly increase the scanning rate of the generation of ultrasound. [4, 5] In this technique, a high power continuous wave (cw) laser beam is scanned across the surface of a material using rotating polygon mirror. The rapid deposit of energy produced by the laser produces thermoelastic expansion along the path on the surface, giving rise to an ultrasonic wavefront in the material. As with pulsed laser generation of ultrasound, sensing of this wavefront will provide information on the material's strength and the location of defects.

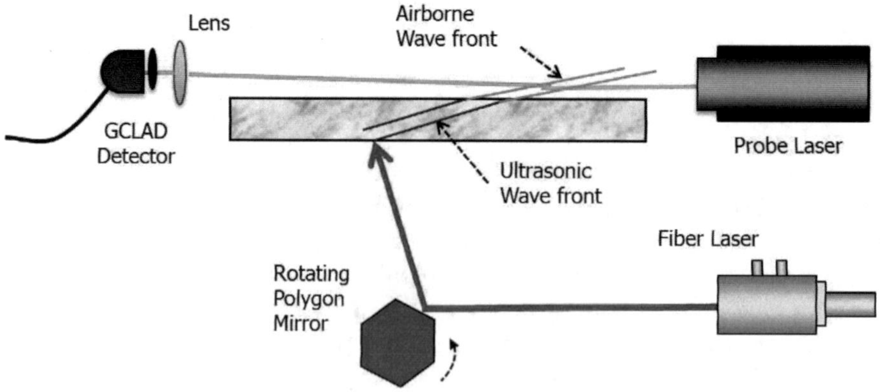

Figure 7. Concept for using GCLAD to detect ultrasound produced by CLGU. The high power cw laser, directed by a fast rotating polygon mirror, creates an ultrasonic wavefront in the material. As the wavefront reaches the solid/air boundary, a portion of the wavefront is radiated into the air and changes the path of the probe beam.

A state-of-the-art pulsed laser with adequate energy and pulse width produces 400 shots per second. To scan a meter line with millimeter resolution would take 2.5 seconds. In contrast, a CLGU system would generate a wavefront along the same line in 0.0003 seconds. The rapid speed is necessary to deposit energy fast enough to create ultrasound in the MHz region. [5] The calculations in reference [4] show that a 10 kW cw laser with a spot size of 2 mm and a scanning rate of 3.05 km/s will produce ultrasound with a frequency of 1.52 MHz.

Detection of this wavefront is an additional challenge. An array of ultrasound transducers can be used to verify the creation of the wavefront, but would be impractical to use for a large scale system. A laser interferometer could also be created that could scan as fast as the generation. A more practical detection system can be made using GCLAD, as depicted in Figure 7. Acting as a line sensor, GCLAD can be positioned such that the probe beam runs parallel to the generation line. The ultrasonic wavefront would radiate into the air and displace the probe beam continuously. A defect in the material would create a break in the wavefront, which would upset the displacement of the beam. The timing of the upset would reveal the location of the defect.

References

[1] Caron J N, Yang Y, Mehl J B, and Steiner K V, 2001 *Materials Evaluation*, **58**(5), p. 667.
[2] Caron J N, 2008 *1st International Symposium on Laser Ultrasonics: Science, Technology and Applications.*
[3] Caron J N, "Continuous laser generation of ultrasound," U.S. Patent No. 8,210,045. 3 Jul. 2012.
[4] Caron J N, Dicomo G P, and Nikitin S, 2011 *Review of Progress in QNDE*, ed D.O. Thompson and D.E. Chimenti, AIP Conference Proceedings, **1430**, p. 243.
[5] Caron J N, Dicomo G P, and Nikitin S, 2012 *Optics letters* **37**(5), p. 830.
[6] Jones T G, Hornstein M K, Ting A C, and Wilkes Z W, 2009 *The Journal of the Acoustical Society of America*, **125**, p. 2556.
[7] Jones T G, Helle M, Ting A C, Kaganovich D, Nicholas M, and Caron J N, 2011, Directed Energy Symposium, La Jolla, CA, p. 14.
[8] Caron J N, 2008 *Review of Progress in QNDE*, ed D.O. Thompson and D.E. Chimenti, AIP, **27**, p.247.
[9] SONOTRONICS, INC., 3169 S. Chrysler Avenue, Tucson, AZ 85713.
[10] Culjat M O, Goldenberg D, Tewari P, and Singh R S, 2010 *Ultrasound in medicine and biology* **36**(6), p. 861.
[11] Oraevsky A A, Jacques S L, and Tittel F K, 1997 *Applied Optics* **36**(1), p. 402.
[12] Zhou S, Reynolds P, Krause R, Buma T, O'Donnell M, and Hossack J A, *IEEE Transactions on Ultrasonics, Ferroelectrics and Frequency Control*, **51**(9), p. 1178.
[13] Kolkman R G M, Blomme E, Cool T, Bilcke M, van Leeuwen T G, Steenbergen W, Grimbergen K A, and den Heeten G J, 2010 *Journal of biomedical optics* **15**(5), p. 055011.
[14] Rousseau G, Blouin A, and Monchalin J-P, 2012 *Biomedical optics express* **3**(1), p. 16.
[15] Wang Y, Li C, and Wang R K, 2011 *Optics letters* **36**(20), p. 3975.
[16] Johnson J L, 2013 "Biomedical Photoacoustic Imaging Using Gas-coupled Laser Acoustic Detection." Student Research Initiative, Boise State University.
[17] Bude R O and Adler R S, *Journal of clinical ultrasound* **23**(4), p. 271.

Journal of Physics: Conference Series **520** (2014) 012023 doi:10.1088/1742-6596/520/1/012023

Evaluation of the Thickness and Bond Quality of Three-Layered Media using Zero-Group-Velocity Lamb Waves

H Cho[1], Y Hara[1] and T Matsuo[1,2]

[1] School of Science and Engineering, Aoyama Gakuin University, 5-10-1 Fuchinobe, Chuo-ku, Sagamihara, Kanagawa, Japan, 252-5258
[2] School of Science and Engineering, Meiji University, 1-1-1 Higashimita, Tama-ku, Kawasaki, Kanagawa, Japan

Email; cho@me.aoyama.ac.jp

Abstract. Bonding conditions such as jointing agent thickness and bond strength (quality) at the interface between two components is a critical concern in an assembled product because the jointing area is often the weakest component in products. Using a laser ultrasonic technique, zero-group-velocity (ZGV) Lamb waves were utilized to characterize the thickness of an adhesive layer for an epoxy-bonded sample and the bond quality for brazed samples. Using two modes of the ZGV Lamb waves, the thickness of the adhesive layer ranging from 0.2 to 0.8 mm can be estimated with high accuracy. In bond quality measurements, poor bond quality lowered the frequency of ZGV Lamb waves. The decrease in the frequency depended on the amount of shear stress along the interface caused by ZGV Lamb waves.

1. Introduction

There are many adhesive components in machines for connecting two different components. It is important to estimate the bonding conditions such as the thickness of the layer and bond strength (quality) for assuring the integrity of the machines. Ultrasonic inspection is a promising method for characterizing bonding conditions. Lamb waves, which propagate in a plate structure, were used for detecting defects at an interface and succeeded in detecting a relatively large defect compared to wavelength of Lamb waves [1]. Reflection waves at interfaces between an adhesive and adherents owing to acoustic impedance mismatch have also been utilized [2]. Frequency dependency of the reflection coefficient revealed the thickness and elastic properties of the adhesive layer. Recently, some researchers have reported that zero-group-velocity (ZGV) Lamb waves have the potential to characterize properties of a layered media [3-5]. ZGV Lamb waves can reveal the elastic properties of anisotropic materials and thickness of a very thin surface layer with high accuracy and high spatial resolution. In this study, ZGV Lamb waves were used to characterize the thickness of an adhesive epoxy layer sandwiched between different materials and the bond quality for brazed material using a laser ultrasonic technique.

2. Zero-group-velocity Lamb waves and experimental setup

Lamb waves are elastic waves that propagate in a plate and have two major modes, symmetric (S) and anitsymmetric (A) modes, in addition to higher modes. Those modes show characteristic dispersions. Figure 1 shows the dispersion relationship of the sample consisting of an epoxy agent (Young's modulus ***E*** = 3.88 GPa, Poisson's ratio ν = 0.38, Density ρ = 1.09 × 10^3 kg/m^3), an aluminum plate (***E*** = 69.0 GPa, ν = 0.335, ρ = 2.8 × 10^3 kg/cm^3, thickness: 3 mm) and an acrylic plate (***E*** = 2.00 GPa, ν = 0.38, ρ = 1.09 × 10^3 kg/m^3, thickness: 1 mm) calculated using Disperse software [6]. The phase velocity of each wave can be calculated by dividing the wavenumber k by the angular frequency ω, while the group velocity corresponds to the gradient of the curve indicating $dk/d\omega$. At the points indicated by the three arrows, the gradient of the curve is zero, which indicates that the group velocity is zero while the phase velocity shows a finite value. These waves are called ZGV Lamb waves. At those points, the wave remains at the area where the wave was generated and creates a resonance vibration. The resonance frequencies of ZGV Lamb waves reflect the local elastic properties, density, and thickness of each plate.

We prepared two different types of samples for estimating the thickness of the bond layer and bond quality. For estimating the thickness, a sample consisting of a 3-mm-thick aluminum alloy plate and a 1-mm-thick acrylic plate glued with an epoxy agent was assembled. The acrylic plate was used for visually checking the bonding condition. The thickness of the epoxy adhesive layer is intentionally changed to estimate the effect of the thickness on the resonance frequency of ZGV Lamb waves. The thickness of the adhesive layer was also measured by subtracting the thickness of the acrylic and aluminum plates from the thickness of the sample with a micrometer. On the other hand, for estimating the bond quality, a sample consisting of an AISI 304 stainless steel plate and a carbon steel plate brazed with Ni amorphous film was fabricated. One sample was treated at the proper brazing temperature (950 °C), and another was treated below the proper temperature (800 °C). To measure adhesive strength with tensile testing, the samples were lap-jointed and the brazed area was 400 mm^2.

Figure 2 shows the experimental setup for measuring the ZGV Lamb waves. A Q-switched pulse YAG laser with a half-duration time of 10 ns, laser spot diameter of 3 mm, and wavelength of 1064 nm was used for generating the ZGV Lamb waves. The energy of the laser pulse was controlled within the thermoelastic region for each sample. The generated ZGV Lamb waves were detected with a laser interferometer with spot size of less than 0.3 mm at the epicenter of the laser irradiation point. The signal from the interferometer was captured with an oscilloscope under 32 time averaging. The samples were set on a two-dimensional translation stage and scanned.

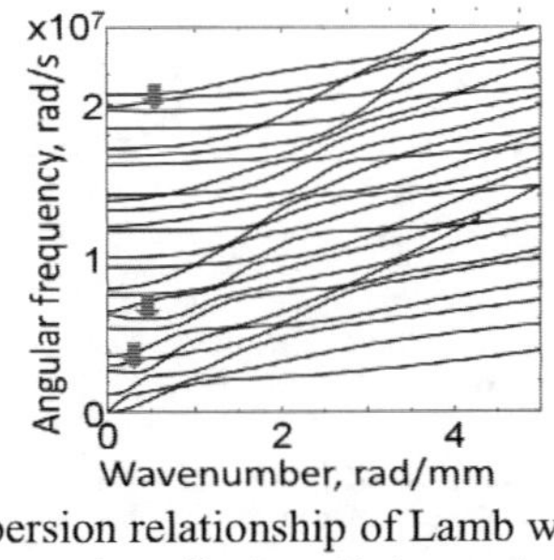

Fig. 1 Dispersion relationship of Lamb waves in three layered media (acrylic/epoxy/steel)

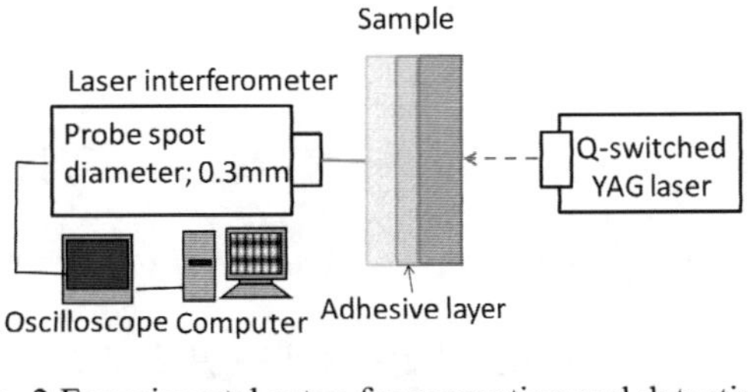

Fig. 2 Experimental setup for generating and detecting ZGV Lamb waves.

3. Results

3.1 Measurement of adhesive thickness

Figure 3 shows the detected waveform and its frequency spectrum for the three-layered sample with an adhesive thickness of 0.25 mm. A thin gold film by DC sputtering was deposited on the acrylic surface to collect a reflection probe beam at the surface. A large pulse corresponding to longitudinal and/or A mode Lamb waves appeared at 30 to 90 μs. The following small oscillating waves corresponded to ZGV Lamb waves. The spectrum shown in Fig. 3 (b) was obtained through a fast Fourier transform of Fig. 3 (a) after a zero padding process. The frequency resolution was improved to 3.84 kHz. Two distinct peaks in the spectrum were observed at approximately 1 MHz and 3.1 MHz and were almost identical to the two higher frequencies of ZGV Lamb waves shown in Fig.1. Figure 4 shows the dependence of two frequencies of measured ZGV Lamb waves on the thickness of the adhesive layer. The frequency of each mode decreased monotonically with an increase in thickness.

The frequency in the high-frequency mode showed a plateau at a thickness less than 0.18 mm and approached the frequency of the S_4 mode ZGV Lamb wave for the substrate aluminum alloy plate (3.1076 MHz). While the peak frequency of the low-frequency mode decreased linearly and at the thickness of less than 0.2 mm, it is difficult to identify the peak owing to its low amplitude. The lines shown in each graph were obtained with the least-squares approximation method. A linear function was selected for the low-frequency mode and a cubic function was selected for the high-frequency mode. The average deviations from the approximation curves were 8 kHz and 9 kHz for the low- and high-frequency modes, respectively, and the estimation errors for thickness were 12 μm and 8 μm for the low- and high-frequency modes, respectively.

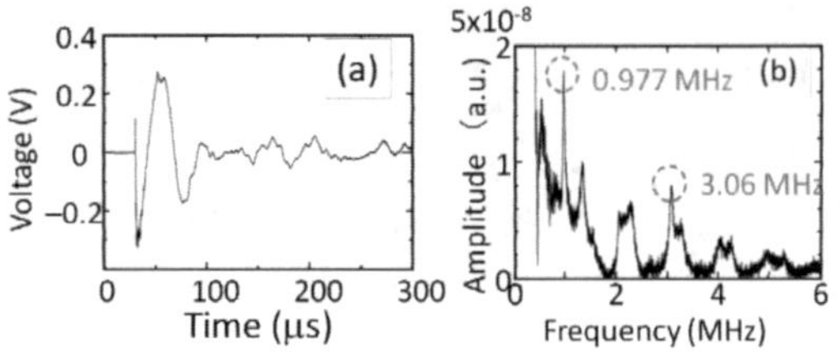

Fig. 3 Waveform (a) and its spectrum (b) for the bonded sample with an epoxy adhesive.

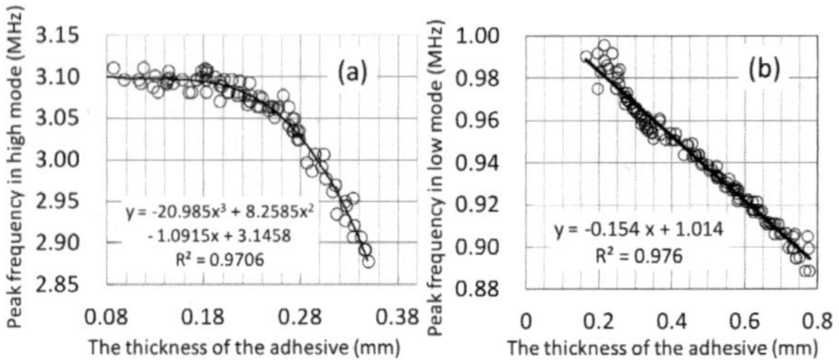

Fig. 4 Relationship between frequency of ZGV Lamb waves and thickness of epoxy adhesive layer.(a) High-frequency mode, (b) Low-frequency mode

3.2 Estimation of bond quality of brazed sample

Figure 5 shows optical microscopy images of cross sections of well- and weak-bonded samples. The thickness of the brazed layer, 80 μm, was almost the same between the well- and weak-bonded samples. In the well-bonded sample, no defect was observed at the interface with the brazed layer, while, black lines at the interface and in the layer can be found in the weak-bonded sample. Those lines correspond to small cracks and imply that the bond quality is poor.

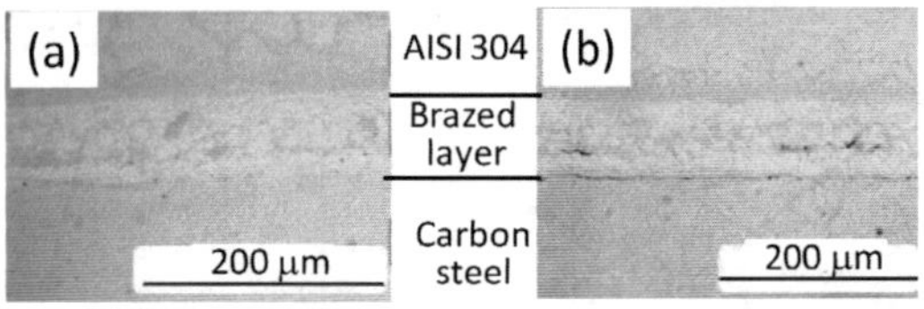

Fig. 5 Cross section of well- (a) and weak- (b) bonded samples.

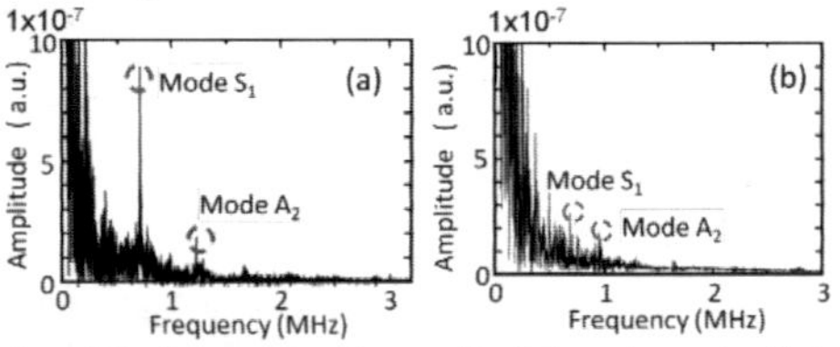

Fig. 6 Amplitude spectra for well- (a) and weak- (b) bonded samples.

Figure 6 shows the amplitude spectra of detected waves obtained with the same experimental setup shown in Fig. 3 for well- and weak-bonded samples. For the well-bonded sample, two distinct peaks at approximately 0.7 MHz and 1.25 MHz appeared. For the weak-bonded sample, two smaller peaks at the same frequency of the two peaks in the well-bonded sample can be found in Fig. 6 (b). To identify the mode for each peak, the dispersion relation for a model that has continuity in displacement and stress through interfaces in the brazed layer was calculated. For the calculation, the Young's modulus, Poisson's ratio, and density of the brazed layer were assumed to be 214 GPa, 0.336, and 5.8 $\times$ 10^3 kg/m^3, respectively [7]. The frequencies of the S_1 and A_2 mode ZGV Lamb waves were calculated as 0.708 MHz and 1.17 MHz, respectively, and were almost equal to the measured peak frequencies. Figure 7 shows the relationship between the peak frequency and amplitude of S_1 and A_2 mode ZGV Lamb waves for well- and weak-bonded samples. Open and closed symbols show the results for the well- and weak-bonded samples, respectively. Circles and squares indicate S_1 and A_2 modes, respectively. The peak frequencies for each mode decreased in the case of the weak-bonded sample, as expected. For the case where the defects are on the interface, the discontinuity in the displacement across the interface is expected, and stiffness at the interface could be reduced [8]. The decrease in the peak frequency of the ZGV Lamb waves could be caused by the reduction in the stiffness. The decrease in the frequency in the A_2 mode was larger than that in the S_1 mode. This implies that the A_2 mode is sensitive to the bond quality. Fig. 8 shows the relative shear stress profiles parallel to an interface between the brazing layer and each substrate through the thickness of the sample for the S_1 and A_2 modes. The stress was normalized with the maximum value in the A_2 mode. The A_2 mode exhibited a high shear stress than the S_1 mode; this implies that the A_2 mode is more sensitive to the bond quality than the S_1 mode in this case. On the other hand, because the peak amplitude of the ZGV Lamb waves in the A_2 mode is almost the same for the well- and weak-bonded sample, and the peak amplitude in the S_1 mode for the well-bonded sample had large dispersion owing to its surface condition, it is difficult to estimate the bond quality of the sample with the amplitude. A tensile test for shear strength for the two brazed samples was conducted using lap-jointed samples. The maximum load for the well-bonded sample (9.13 kN) was much larger than that in the weak-bonded sample (3.91 kN). This result supports the conjecture that a decrease in the peak frequency of ZGV Lamb waves indicates poor bond quality.

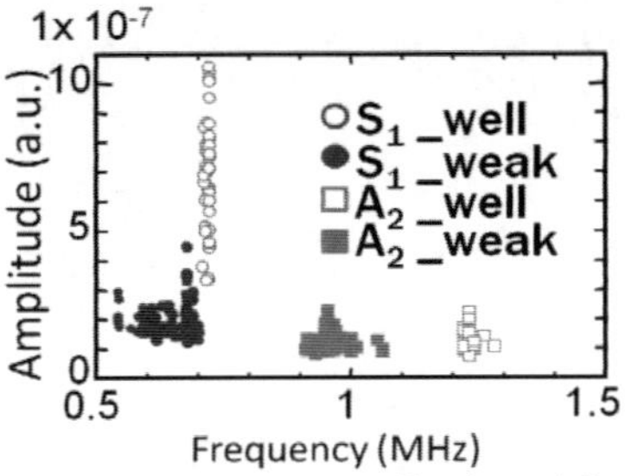

Fig. 7 Relationship between amplitude and frequency of ZGV Lamb waves for the well- and weak-bonded sample.

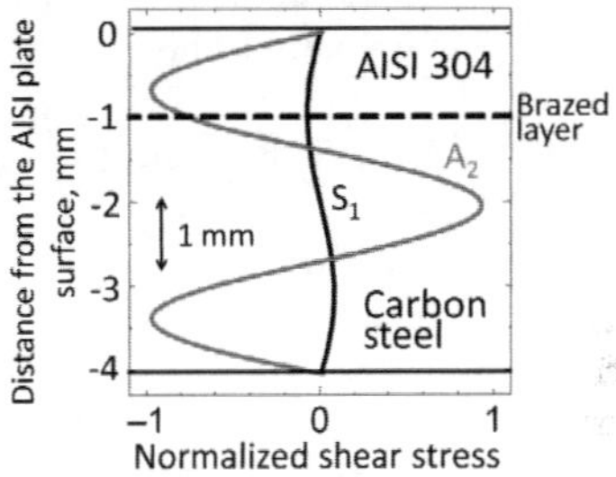

Fig. 8 Shear stress profiles through the thickness of the sample for S_1 and A_2 mode ZGV Lamb waves.

4. Conclusions

The thickness of the adhesive layer for an epoxy-bonded sample and bond quality for a brazed sample was estimated by examining the peak frequency of zero-group-velocity (ZGV) Lamb waves.

In thickness measurements of an adhesive layer in the epoxy-bonded sample, two distinct peaks at around 0.95 MHz and 3 MHz that corresponded to the ZGV Lamb waves were observed. Those peak frequencies monotonically decreased with an increase in the thickness of the adhesive. Average deviations from the approximation curves obtained from the measured peak frequencies were as low as 12 μm and 8 μm for low- and high-frequency modes, respectively.

In the estimation of bond quality in the brazed sample, two peaks appeared at 0.7 MHz and 1.25 MHz, and corresponded, respectively, to S_1 and A_2 mode ZGV Lamb waves. The peak frequencies shifted to a lower frequency for the weak-bonded sample. The decrease in peak frequency of the A_2 mode was larger than that in the S_1 mode for the weak-bonded sample. This difference in the decrease for the weak-bonded sample can be explained with shear stress at the interface on the brazed layer.

Acknowledgement

This study was supported by a JSPS Grant-in-Aid for Scientific Research (C) (25420031).

References

[1] Ramadas C, Padiyar M J, Balasubramaniam K, Joshi M and Krishunamurthy C V, 2010 *J Intelligent Material Systems and Structure* **21** 817
[2] Vijaya kumar R L, Bhat M R and Murthy C R L, 2013 *Ultrasonics* **53** 1150
[3] Clorennec D, Prada C and Royer D, 2010 *IEEE Tran on UFFC* **57** 1125
[4] Prada C, Clorennec D, Murray T W and Royer D, 2009 *J Acoust Soc Am* **126** 620
[5] Holland S D and Chimenti D E, 2004 *Ultrasonics* **42** 957
[6] Pavlakovic B and Lowe M, 2005 DISPERSE software, v.2.0.6.
[7] Cho H, Futatugi T, Ogawa S and Takemoto M, 1996 *NDT&E International* **29** 301
[8] Schoenberg M, 1980 *J Acoust Soc Am* **68** 1516

Numerical Analysis of Narrow Band Ultrasonic Wave Generation with High Repetition Pulse Laser and Laser Scanning

T Hayashi, K Yamaguchi and S Biwa
Graduate School of Engineering, Kyoto University, Kyoto, 615-8540, Japan

E-mail: hayashi@kuaero.kyoto-u.ac.jp

Abstract. Although the easiest way to enhance ultrasonic energy generated with pulse laser is to increase laser output, excessive laser output causes damage of the surface. This study introduced an alternative way to generate burst signals without any damages at the surface using a newly developed high repetition pulse laser controlled by galvano mirrors. The calculation results using two-dimensional elastodynamic finite integration technique coupled with thermoelastic effect proved that burst wave of 1 MHz and its higher harmonics were generated while supressing excessive temperature rise using this technique. Moreover, significantly large displacements at the frequency range sufficiently lower than laser repetition rate were observed of the same order of displacements generated with one single shot with the same input energy.

1. Introduction

In laser ultrasonics, ultrasonic generation is feasible with pulsed laser emission, while ultrasonic detection requires laser beam reflected or scattered from a surface of the object. Therefore, ultrasonic detection by laser may have some problems when applying it to alfresco civil structures and products in a production line due to their difficult conditions such as roughness, dirt, rust, and paint of surface and large vibration with a low frequency. Thus, authors adopted a scanning laser source technique (SLS) to solve such practical problems and have developed defect-imaging technique using the SLS for the purpose of application to such alfresco structures and in-line inspection.

In the SLS, ultrasonic waves are generated with pulsed laser and detected with ultrasonic transducers fixed on the surface of an object and the laser source is scanned over the surface with galvano mirrors. Authors proved that an amplitude distribution obtained by the SLS corresponds to an image of defects in a thin plate and that clearer images can be obtained using multiple receiving transducers. Moreover, authors showed that we can replace the receiving transducers by non-contact air-coupled transducers when using a low-frequency range.

Although we realized the full non-contact technique, its application is limited due to its low signal to noise ratio. This study, therefore, introduces burst wave generation technique using high repetition pulse laser and investigates the effectiveness of the technique with numerical calculations.

2. Burst wave generation using a high repetition pulse laser controlled by galvano mirrors

The easiest way to enhance ultrasonic energy generated with pulse laser is to increase laser output. However, as excessive laser output causes damage of the surface, it cannot be applied to structures and products whose surfaces should be kept undamaged. We now investigate the burst wave generation without any damages of a surface using a newly developed high repetition pulse laser, in which laser pulses move at small distance with galvano mirrors to avoid excessive temperature rise.

Hereafter, we evaluate the burst wave generation technique with the numerical calculations of wave propagation and temperature change which is conducted for an aluminium alloy plate of 0.1 mm thickness using a two-dimensional elastodynamic finite integration technique (EFIT) coupled with thermoelastic effect [1]. As optical penetration of laser emission to aluminium is negligibly small compared to the spatial mesh division, heat is supplied at the infinitesimally thin layer of the plate surface $y = 0$ as,

$$Q(x,y,t) = \sum_{n=1}^{N} Q_n(x,y,t)$$

$$Q_n(x,y,t) = \begin{cases} \dfrac{4q_0}{\sqrt{\pi} d_0 t_0} \dfrac{(t-T_n)}{t_0} \exp\left(-\dfrac{t-T_n}{t_0}\right) \exp\left(-\dfrac{4(x-X_n)^2}{d_0{}^2}\right) \delta(y) & t \geq T_n \\ 0 & t < T_n \end{cases} \quad (1)$$

where N is the number of laser shots, Q_n is the heat supply for the nth shot, T_n and X_n are temporal and spatial shift for the nth shot, d_0 and t_0 are the width of laser source and rise time, q_0 is the total heat supply per unit length for a single shot[2], [3]. T_n and X_n in Eq. (1) are described using temporal increment Δt and spatial increment Δx as,

$$T_n = (n-1)\Delta t\,, \quad X_n = (n-1)\Delta x\,. \quad (2)$$

The area integration of the heat supply expressed by equation (1) in EFIT formalization results in the line integration on the plate surface $y = 0$. All parameters used in the calculation are listed in Table 1.

Figure 1 represents a schematic figure of the burst wave generation with a pulse laser and laser scanning. The A0 mode of Lamb wave generated in the frequency-thickness product range of 1 MHz x 0.1 mm have the phase velocity of about 1000 m/s and the wavelength of about 1 mm. As the spatial interval of scanning Δx is sufficiently small compared to the wavelength, from the view point of ultrasonic propagation, ultrasonic wave is generated at approximately the same positions. In general, the product of spatial interval Δx and the number of laser shots N should be smaller than half wavelength for effective burst wave generation, where N corresponds to the number of burst cycles to generate. Namely, this relationship determines the effective frequency range in this technique.

Figure 2 shows the temperature change at the laser spots: (a) is for a single shot ($N = 1$), (b) is for 10 shots at the same position ($N = 10$, $\Delta x = 0$), and (c) is temperature change at the fifth laser spot for 10 shots with scanning ($N = 10$, $\Delta x = 40$ μm). As the temperature change is linearly proportional to the total energy input q_0, the vertical axis in figure 2 is represented by the temperature divided by q_0. In order to compare (a), (b) and (c) under the same input energy, the temperature change for the input

Table 1. Parameters used in the calculations.

<table>
<tr><td rowspan="4">Material

(Aluminium plate)</td><td colspan="2">Cross-section size</td><td colspan="2">Density</td><td>Longitudinal wave velocity</td><td colspan="2">Transverse wave velocity</td></tr>
<tr><td colspan="2">10 0mm x 0.1mm</td><td colspan="2">2700 kg/m^3</td><td>6400 m/s</td><td colspan="2">3200 m/s</td></tr>
<tr><td colspan="2">Specific heat</td><td colspan="2">Thermoelastic coupling constant</td><td>Thermal conductivity</td><td colspan="2">Base temperature</td></tr>
<tr><td colspan="2">900 J/kg/K</td><td colspan="2">5.6 x10^{-3} K^{-1}</td><td>160 W/m/K</td><td colspan="2">300 K</td></tr>
<tr><td rowspan="2">Temporal and spatial mesh division</td><td colspan="2">Temporal division</td><td colspan="2">Spatial division</td><td>Number of divisions</td><td colspan="2">Number of temporal iteration</td></tr>
<tr><td colspan="2">1 ns</td><td colspan="2">0.01 mm (both directions)</td><td>10000 x 10</td><td colspan="2">10000 (up to10μs)</td></tr>
<tr><td rowspan="2">Single line source parameters</td><td colspan="3">Width of line source (d_0)</td><td colspan="2">Rise time (t_0)</td><td colspan="2">Total energy input (q_0)</td></tr>
<tr><td colspan="3">200 μm</td><td colspan="2">10 ns</td><td colspan="2">q_0 J/m</td></tr>
<tr><td rowspan="2">Firing repetition parameters</td><td colspan="3">Repetition frequency (=1/ Δt)</td><td colspan="2">Repetition time interval (Δt)</td><td colspan="2">Spatial interval (Δx)</td></tr>
<tr><td colspan="3">1 MHz</td><td colspan="2">1 μs</td><td colspan="2">40 μm</td></tr>
</table>

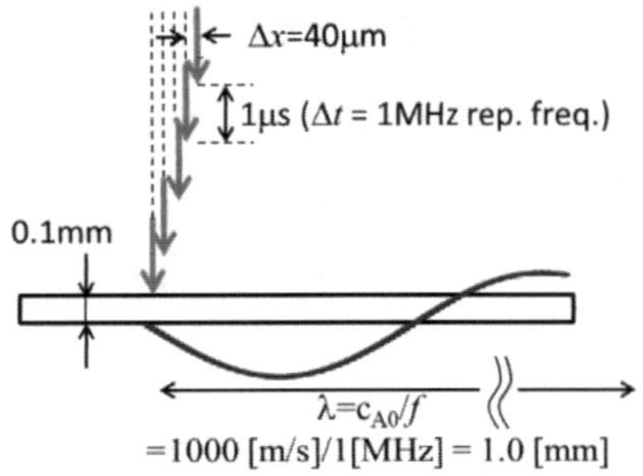

Figure 1. Schematic figure of burst wave generation with a pulse laser controlled by galvano mirrors.

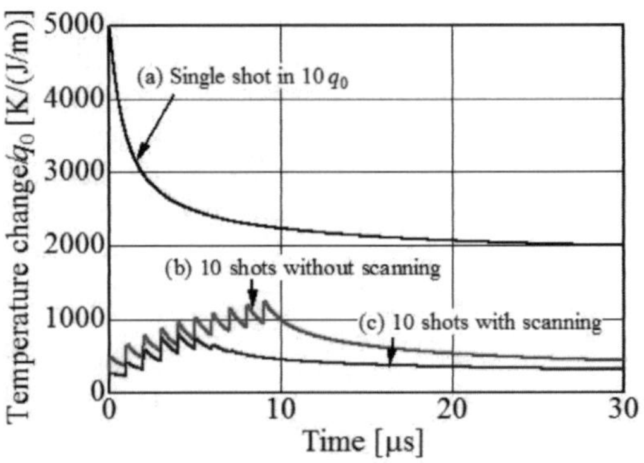

Figure 2. Temperature change at the laser spot.

energy of $10q_0$ is shown in (a). Comparing the maximum value of (c) with those of (a) and (b), we can find that the temperature rise could be suppressed about 1/6 of that for the single shot (a) and about 2/3 of that for 10 shots with no scanning (b) by the use of a high repetition laser and scanning. The temperature rise in (c) is approximately determined by the number of times laser is incident on the measurement point of temperature. Therefore, when the spatial interval of laser scanning Δx is larger than the width of line source d_0 , the maximum temperature of the surface becomes about 1/10 of that for the single shot.

Next, temporal displacements at 20 mm from the first laser source (n=1) and their frequency spectra are shown in figure 3. In (a), significantly large chirp-like waveform was observed as a natural vibration of A0 mode of Lamb wave generated in the aluminium thin plate with the thickness of 0.1 mm. As the group velocity of the A0 mode monotonically increases with frequency in the range of frequency- thickness product, the signal period gradually increase with time. In (b) and (c), waveforms between about 8 μs and 20 μs are signals controlled by the multiple laser emission having components mainly about 1 MHz. The waveforms of (b) and (c) are very similar, which proves that the laser scanning does not affect significantly the temporal displacements. In the frequency spectra of (b) and (c), main peaks were obtained about at 1 MHz as well as the higher harmonics. Comparing the peak positions of (b) and (c), (c) provided frequency peaks at slightly higher frequency than (b), which is due to the Doppler shift occurred by moving laser spots in Δx .

Since large displacements with low frequency components were observed in figure 3 (a) as well as (b) and (c) after 20 μs, figure 4 shows extended waveforms of figures 3 (a) - (c) until 100 μs and their frequency spectra. These three waveforms after 60 μs are similar except for small vibration with high frequency in (a) after 70 μs which is the reflected waves from the plate end. These results represent that waveforms in low frequency range were not affected by the number of laser shots because the repetition rate is sufficiently high compared to the low frequency oscillation.

3. Visualization of temperature change and wave propagation

Figure 5 is the visualization results of the region of 2.0 x 0.1 mm^2 around laser spots: (a) is temperature change, and (b) and (c) are wave propagations showing displacement data magnified appropriately. Since the generated wave is dominated by the low frequency components as shown in (b), in order to visualize the targeting burst wave of 1 MHz, (c) is obtained by filtering with the high-pass filter of 0.6 MHz cut-off frequency. We can recognize that the laser shots deform the thin plate in (b) and that ultrasonic burst wave is generated in (c).

4. Conclusions

This study presented the results of numerical calculation for the burst wave generation technique with the combination of high repetition laser emission and laser scanning. The calculation results proved that burst wave of 1 MHz and its higher harmonics can be generated while suppressing excessive temperature rise using this technique. Moreover, significantly large displacements at the frequency range sufficiently lower than laser repetition rate were observed of the same order of displacements generated with one single shot with the same input energy.

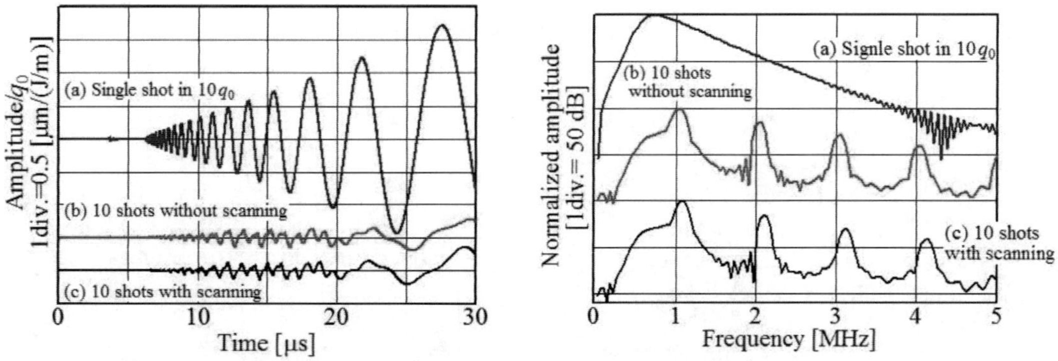

Figure 3. Displacements in the thickness direction and their frequency spectra for $q_0 = 1$ at 20 mm from the first laser spot.

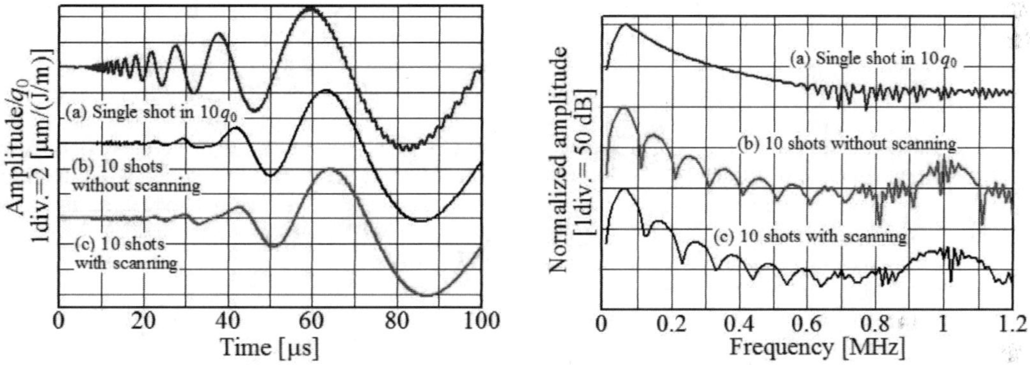

Figure 4. Extended waveforms of figure 3 and their frequency spectra

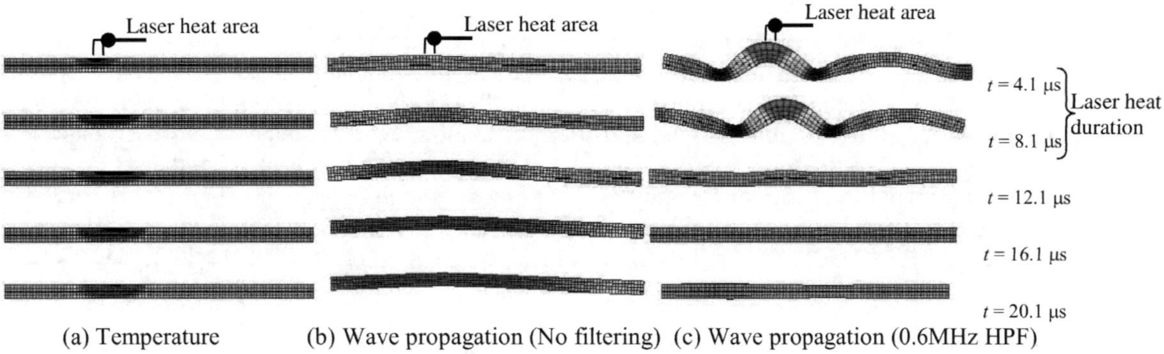

Figure 5. Visualization of calculation results for N=10 with scanning.

5. References

[1] Yamaguchi K, Biwa S and Hayashi T 2013 *Proc. of ICU 2013* 518 - 9
[2] Arias I and Achenbach J D 2003 *Int. J. Solids Struct.* **40** 6917 - 35
[3] Sun H and Zhang S 2010 *J. Appl. Phys.* **108** 123101

Journal of Physics: Conference Series **520** (2014) 012025 doi:10.1088/1742-6596/520/1/012025

Development of Polygonal Buffer Rods for Ultrasonic Pulse-Echo Measurements

F M Foudzi[1] and I Ihara[2]
[1]Graduate School of Nagaoka University of Technology, Nagaoka 940-2188, Japan
[2]Department of Mechanical Engineering, Nagaoka University of Technology, Nagaoka, Niigata 940-2188, Japan

Email: ihara@mech.nagaokaut.ac.jp

Abstract. In ultrasonic pulse-echo measurement with a long buffer rod (waveguide), it is required to prevent the generation of spurious echoes (often called trailing echoes) accompanying with a main echo in the buffer rod. In this work, new method to prevent such trailing echoes in the rod is proposed and the effectiveness of the method has been demonstrated experimentally and numerically. In the method the cross-sectional shape of the rod perpendicular to the axial direction is a polygon having sides any one of which is not parallel to any of the other sides, so that trailing echoes are hardly generated during the propagation of pulsed ultrasonic wave in the rod. Three-dimensional numerical simulations based on a finite different method are performed to examine the behaviours of ultrasonic pulse-echoes including trailing echoes for several types of buffer rods having different cross-sectional shapes such as a circle, triangle, square, pentagon, hexagon and heptagon. Based on the results, experiments with several buffer rods are carried out at frequency 5 MHz. It has been found that heptagon may be the suitable shape for effectively eliminating trailing echoes and improving the signal-to-noise ratio of the measured pulse-echo.

1. Introduction

In the several fields of science and engineering, ultrasonic pulse-echo techniques are widely used for materials evaluations and characterizations. It is known for the applications under harsh conditions such as high temperature environments that some high temperature ultrasonic techniques such as laser ultrasonics, EMATs, high temperature transducers and buffer rod method (known as a delay-line or waveguide method) can be available. It is noted that each technique has advantages and disadvantages. Although the buffer rod method is rather classical, it is still an attractive approach from the viewpoints of its simplicity, convenience, robustness and low cost. For very high temperature applications of the buffer rod method, a long buffer rod is often employed as a waveguide. A conventional piezoelectric transducer is installed to the one end of the buffer rod and the other end is in contact with a measuring objective at high temperature. A crucial problem in using such long buffer rods is, in most cases, caused by spurious echoes (known as trailing echoes) due to interference of mode converted waves, dispersion and diffraction within the rod of finite diameter [1, 3, 4]. Such spurious echoes are considered as noise and overlapped with a main echo while pulse-echo measurement is being performed and deteriorate the signal-to-noise ratio (SNR) in the measurement significantly because of their possible interference with a desired signal. To overcome such problem, tapered buffer rods were developed to prevent the generation of trailing echoes [1-3]. It should be noted, however, that such tapering is not always the best solution and may have a difficulty in machining if the rod material is "hard-to-work material" such as hard ceramic materials. In this paper, an alternative method to prevent such trailing echoes in long buffer rods is proposed and the effectiveness of the method is demonstrated experimentally and numerically.

2. Designing polygonal buffer rods

It is known that trailing echoes appeared in a long buffer rod is mainly caused by interferences of reflected and mode converted waves at the side surface of the rod during ultrasonic wave propagation through the rod [1, 3, 4]. In the case of a cylinder rod whose cross-sectional shape perpendicular to the axial direction is a circle as shown in the left side of figure 1, there exist infinitely many pairs of reflected or mode converted echoes that will constructively interfere with each other depicted by arrows in the figure. This is basically because of axisymmetric shape of the circle. Such interferences make trailing echoes effectively accompanying with a main echo propagating through the rod. Now, we consider different shapes for the cross-sectional geometry, such as triangle or square as shown in the second or the third from the left in figure 1. In the case of square, because of two pairs of parallel sides, there are possibly two pairs of waves to interfere with each other depicted as arrows in the figure, so that it is still expected to produce trailing echoes in the rod. On the other hand, in the case of triangle, there is no possible interference of waves because there is no pair of parallel sides as shown in the figure. Therefore, it is highly expected for the triangle shape to prevent the generation of trailing echoes, owing to less interference of waves. The same thing is true for pentagon and heptagon but not for hexagon. Thus, a polygon having sides any one of which is not parallel to any of the other sides seems to be effective to reduce the formation of trailing echoes. It is predicted that "odd polygons (triangle, pentagon and heptagon)" can effectively restrict the generation of trailing echoes compared to "even polygons (square and hexagon)".

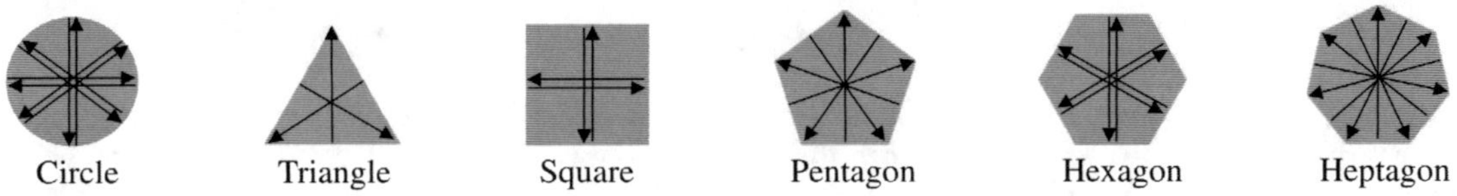

Figure 1. The cross-sectional shapes of polygonal buffer rods. Arrows depicted in each shape show the possible interference waves reflected or mode converted at the side surface of the rod.

3. Numerical simulations

Three-dimensional numerical simulations based on a finite difference method are performed to examine the behaviors of ultrasonic pulse echoes including trailing echoes for several buffer rods with different polygonal shapes such as cylinder, triangle, square, pentagon, hexagon and heptagon. Six kinds of buffer rods models with 25 mm in length and approximately 5 mm in diameter are employed as shown in figure 2. The size of ultrasonic transducer is 2.2 mm in diameter and the operating frequency is 5 MHz. A commercial software, Wave 3000 from CyberLogic, Inc. is used for the simulations and 0.03 mm grid size for each model is used to assure the accuracy and stability in the calculation.

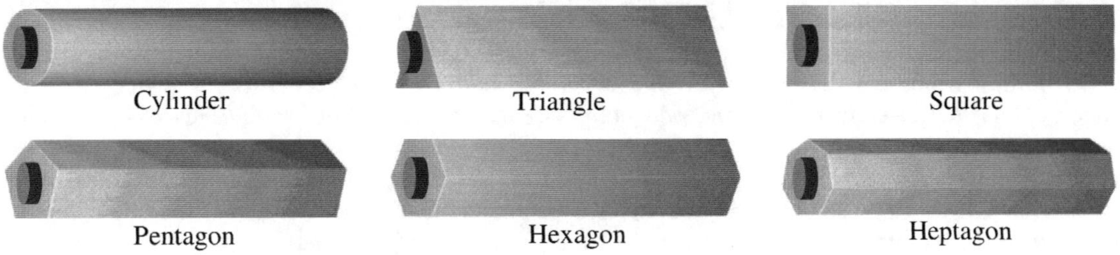

Figure 2. Three-dimensional models of polygonal buffer rods for the numerical simulations.

Figure 3 shows the captured images of ultrasonic pulse echoes propagating through the cylinder, square and pentagon rods, where brighter in the image means higher amplitude in the ultrasonic wave. The cross sectional images at a certain position for each echo are also shown under the side view images. As expected, the generation of a trailing echo accompanying with a main echo (first echo) is clearly seen in the cylinder rod as showin in figure 3(a). It is noted that the distance between the main and trailing ehoes is about 7.8 mm and almost agree with that estimated theoretically [5]. It can be seen in figure 3(b) that a clear trailing echo is also observed for the square rod and the amplitude of the echo is even larger than that of the cylinder rod. This may be caused by more effective interference of reflected and mode converted waves. Thus, the formation of trailing echoes for such rods are succesfully demonstrated with the three-dimensional simulations. On the other hand, no trailing is observed for the pentagon rod as shown in figure 3(c). The similar result is obtained for the heptagon rod. This reveals that the generation of trailing echoes is effectively constrained in "odd polygon" because there is no pair of parallel sides in such odd polygon rods as mentioned in Section 2. This fact is clearly shown in figure 4 showing simulated results of round-trip pulse echoes for all buffer rods.

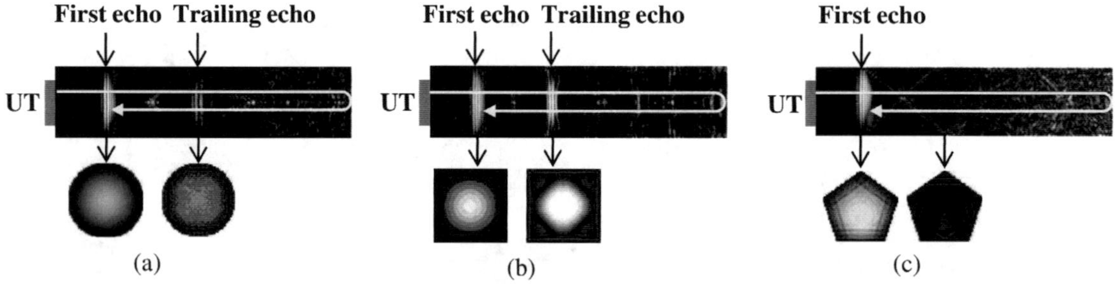

Figure 3. Captured images of pulsed ultrasonic waves propagating through the buffer rods: (a) cylinder, (b) square and (c) pentagon, from the side view (top images) and cross-sectional view (bottom images).

4. Experimental demonstrations

To demonstrate the validity of the numerical simulation results and the effectiveness of the polygonal rods in eliminating trailing echoes, experiments was carried out for cylinder and polygonal rods with 100 mm in length and approximately 20 mm in diameter. The rods in the experiment are similar in shape to those in the numerical simulation but different in size. Main reason of using small size models in the simulation is time-consuming problem in calculation. It has been confirmed in the simulation that the tendency in the wave propagation behaviour including trailing echoes are almost independent from the rod size as long as the shapes are similar. A 5 MHz ultrasonic transducer of 8 mm in diameter was used for the measurement. Figure 5 shows measurement results of round-trip pulse echoes for the polygonal rods. It can be seen from these results that the tendency in the appearance of trailing echoes for these rods almost agree with that of simulation results shown in figure 4. Although there were some discrepancies in the amplitude of trailing echoes, such quantitative difference may be caused by uncertainty in numerical and experimental conditions. It is noteworthy that the trailing echoes were markedly reduced for the odd polygon rods such as pentagon and heptagon. Thus, the effectiveness of the odd polygon rods in eliminating trailing echoes has been demonstrated.

A remarkable trailing echo was observed for the triangle rod as shown in the second from the top of figure 5. This reveals that unexpected interferences occurred in the triangle rod. It should also be noted here that there exists another remarkable echo for the pentagon rod, depicted by a circle in the forth from the top of figure 5. This echo is quite unpredictable because its location of occurrence is far from the theoretical estimation. Since this unexpected echo was appeared at high reproducibility and independent from measurement conditions, it is likely that this echo was generated owing to the

Journal of Physics: Conference Series **520** (2014) 012025 doi:10.1088/1742-6596/520/1/012025

geometrical feature of pentagon. To examine this prediction, three-dimensional numerical simulations was performed for a pentagon rod of the same geometry as used in the experiment. As the result, a similar echo was observed at the same location as the measured one, as depicted by a circle in figure 6. The further studies are needed to examine the formation of such echoes.

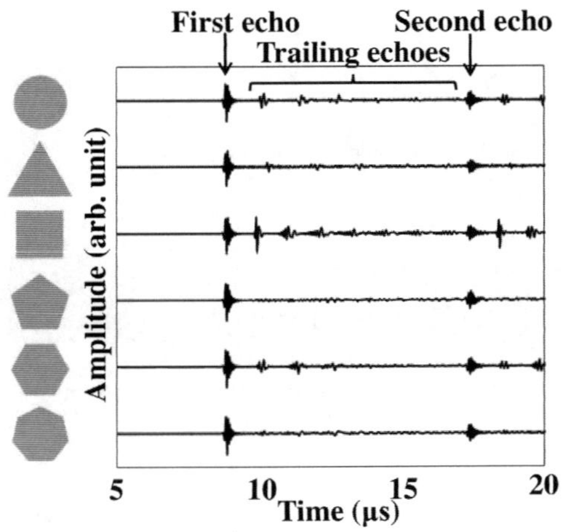

Figure 4. Simulation results of round-trip pulse echoes for polygonal rods.

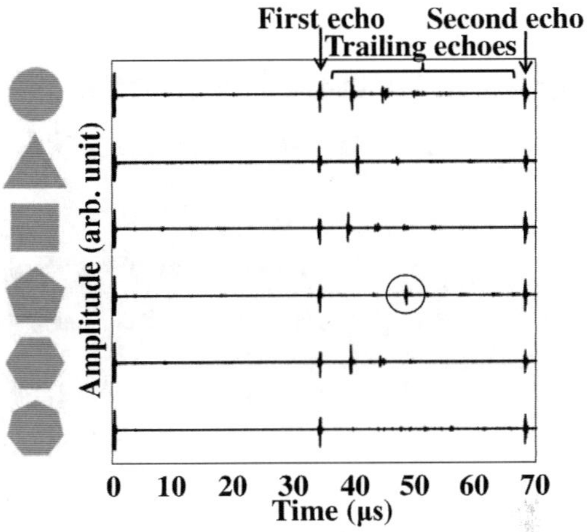

Figure 5. Measured results of round-trip pulse echoes for polygonal rods.

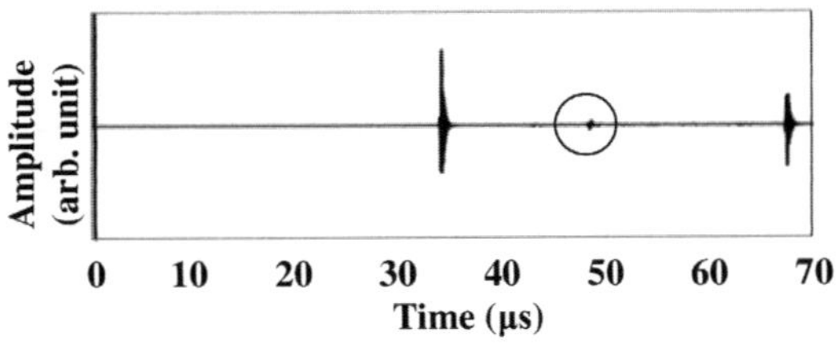

Figure 6. Simulation result of round-trip pulse echoes for a pentagon rod of the same geometry as used in the experiment.

5. Conclusions

A new method to prevent trailing echoes in long buffer rods is proposed and the effectiveness of the method is demonstrated experimentally and numerically. In the method the cross-sectional shape of the rod perpendicular to the axial direction is a polygon having sides any one of which is not parallel to any of the other sides. It has been found that odd polygons such as pentagon or heptagon are suitable shape for effectively eliminating trailing echoes. Further studies are needed to design an optimum buffer rod suitable for pulse echo measurements with a high signal-to-noise ratio.

6. Acknowledgment

Support by the Grant-in-Aid for Scientific Research (B) 25289238 from the JSPS is greatly appreciated.

7. References

[1] Jen C K, Cao B, Nguyen K T, Loong C A and Legoux J G 1997*Ultrasonics* vol 35 p 335
[2] Ihara I, Jen C K and Franca D R 2000 *Japanese Journal of Applied Physics* vol 39 p 3152
[3] Ihara I, Jen C K and Franca D R 2001 *Advanced in Technology of Materials and Material Processing Journal* vol 3 p 45
[4] Jen C K, Piche L and Bussiere J F 1990 *The Journal of the Acoustical Society of America* vol 97 p 23
[5] Redwood M 1960 *Mechanical Waveguides* (New York: Pergamon) chapter 9 pp 191-207

Journal of Physics: Conference Series **520** (2014) 012026 doi:10.1088/1742-6596/520/1/012026

Quantitative Measurement of Highly Focused Ultrasound Pressure Field by Optical Shadowgraph

R Miyasaka[1], S Harigane[1], S Yoshizawa[1] and S Umemura[2]

[1]Department of Electrical and Communication Engineering, Tohoku Univ., Sendai 980-8579, Japan
[2]Department of Biomedical Engineering, Tohoku Univ., Sendai 980-8579, Japan

Email: ryo-m@ecei.tohoku.ac.jp

Abstract. In the development of medical ultrasound techniques, fast and accurate pressure field measurement is important. The most common method to measure an ultrasound pressure field is mechanically scanning a hydrophone, which takes a long time and might disturb the acoustic field. In this study, we used an optical shadowgraph method. To perform this method quantitatively, it is important to define the optical propagation length precisely. For this purpose, a holographic diffuser was used as the imaging screen. Combined with a computed tomography (CT) algorithm, a pressure field was reconstructed, and the result was compared with that of hydrophone measurement. By using two shadowgraph data from short and long propagation lengths, the pressure field was successfully reconstructed even at a pressure level for high intensity focused ultrasound (HIFU) treatment.

1. Introduction

Accurate measurement of ultrasound field is important to ensure and improve the safety and efficacy of the medical use of ultrasound such as HIFU (High Intensity Focused Ultrasound) treatment. Fast and accurate measurement is important to quality-control, plan, and assess such treatment. The most common method to measure an ultrasound pressure field is mechanically scanning a hydrophone, which takes a long time and might disturb the acoustic field. In contrast, an optical method takes very short time and does not disturb the acoustic field.

Optical methods in previous studies [1-6] succeeded in reconstructing an ultrasound field using a computed tomography (CT) algorithm up to several hundreds kPa_{pp}. In the optical shadowgraph method, the optical propagation length between the ultrasound field, where the light receives the optical phase modulation, and the optical imaging plane is important for the accuracy. In this study, we improve this method by precisely defining the optical propagation length through setting a holographic diffuser as the imaging screen at the imaging plane. From the projections of the pressure field, obtained thereby, the three-dimensional (3D) ultrasound pressure field is reconstructed by a CT algorithm.

2. Materials and Methods

2.1 Determination of projected distribution of ultrasound pressure

The spatial change of water density due to acoustic pressure, distort the incident planar optical

3rd International Symposium on Laser Ultrasonics and Advanced Sensing IOP Publishing
Journal of Physics: Conference Series **520** (2014) 012026 doi:10.1088/1742-6596/520/1/012026

wavefront to a wavy one. The light at this moment is considered to have received only phase modulation without intensity modulation. The phase variation modulates the optical intensity during propagation and forms the shadowgraph image. Figure 1 shows the optical intensity modulation in a shadowgraph [3]. The ultrasound and optical propagation directions are defined by y and z, respectively. I_0 is the optical intensity distribution of the light that has just passed through the ultrasound field, and S_0 is the area. I and S are those on the imaging plane. Assuming the optical refraction angle θ_y due to the ultrasound pressure distribution is sufficiently small, the relation between optical intensity and acoustic pressure can be written as follows.

$$\frac{I_{on}-I_{off}}{I_{on}} = -l\frac{\partial n}{\partial p}\left(\frac{\partial^2}{\partial x^2}+\frac{\partial^2}{\partial y^2}\right)\int pdz \qquad (1)$$

Here, the geometrical optics approximation was applied. I_{on} and I_{off} are the optical intensities on the imaging plane with and without ultrasound exposure, respectively. The piezo-optic coefficient [7], $\partial n/\partial p$ for water is calculated to be 1.32×10^{-10} Pa^{-1} from its density of 10^3 kg/m^3, sound speed of 1500 m/s, refractive index of 1.34 at 20 ℃, and the optical wavelength of 589 nm. A two-dimensional Helmholtz equation for the acoustic pressure distribution projected to the x-y plane can be solved in a spatial frequency domain.

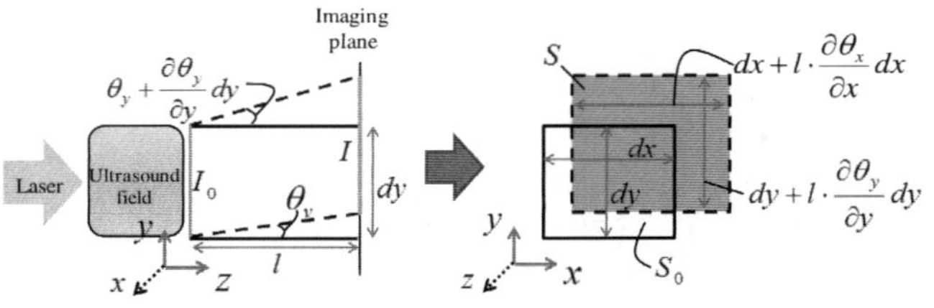

Figure 1. Intensity modulation due to phase variation.

2.2 Estimation of optical propagation length

We need to measure the optical propagation length l in eq. (1) between the imaging plane and the axis of the ultrasound pressure field. In previous study, we measured the optical depth of field and defined its center as the position of the imaging plane. However, because the depth of field is not small enough, the measurement error is relatively high when the optical propagation length l is short. To solve this problem in this study, a holographic diffuser was employed as the imaging screen to precisely define the optical propagation length.

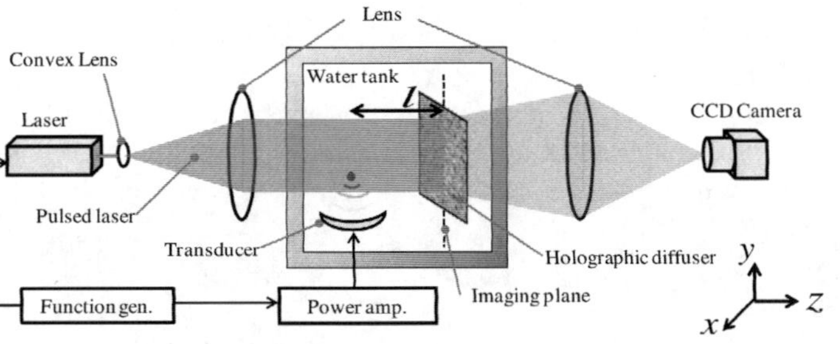

Figure 2. Optical measurement setup.

2.3 Experimental setup

The experimental setup was shown in figure 2. A solid state pulsed laser (CryLas FDSS 532-Q2, wavelength: 532 nm, pulse length: not more than 1.3 ns, power: 4 kW) was used as a light source. The light was expanded by a convex lens (diameter: 3 mm, focal length: 3 mm) and collimated by a lens (diameter: 150 mm, focal length: 1500 mm) in front of the water tank. The light having passed through the ultrasound field was converged by a lens behind the water tank. We measured the optical depth of field and set a holographic diffuser (Optical Solutions, transmittance: above 90 %, diffuse angle: 5

Journal of Physics: Conference Series **520** (2014) 012026 doi:10.1088/1742-6596/520/1/012026

degree) there as the imaging screen. Here, its amplification efficiency is different at each point on the diffuser. For this reason, we moved a holographic diffuser slightly up and down in capturing an enough number images and averaged them. The shadowgraph images were taken by a CCD camera (Sony XCD-U100, 1200 × 1600 pixel). The laser pulse and the ultrasound from the transducer were synchronized by a function generator (NF WF 1974) exciter every 1 ms, and the shutter speed of the CCD camera was 1 ms. A lead zirconate titanate (PZT) transducer (aperture and diameter: 70 mm, center frequency: 1.14 MHz) was used to generate the ultrasound pressure field to be reconstructed. 50 images were acquired and averaged with and without ultrasound exposure, which correspond at I_{on} and I_{off} in eq. (1), respectively. The 2D Helmholtz equation in eq. (1) was numerically solved in a spatial frequency domain for $k^2(=k_x^2+ k_y^2)$, from which the 3D ultrasound pressure field was reconstructed using a CT algorithm. The 3D reconstruction of an ultrasound pressure field generally requires multiple angular projection data, obtained by rotating the transducer 180° along the y-axis. However, owing to the axial symmetry of the transducer, the projection data at only an angle was needed.
We compared the reconstruction from optical measurement and the measurement by a hydrophone (Onda HGL-0085) with an active diameter of 0.085 mm. The hydrophone was scanned every 400 μm on the plane.

3. Results and Discussion

Figures 3 (a) and (b) show the normalized instantaneous pressure field of focused ultrasound generated by the symmetric transducer in the *x-y* plane, reconstructed from optical measurement and measured by a hydrophone, respectively. Here, the optical propagation length was 50 mm, and the peak pressure value was 3 MPa_{pp}. A short optical propagation length was chosen to measure a high pressure. To compare the two methods precisely, the data on the axial direction (*x* axis) and the lateral direction (*y* axis) are shown in figures 4 (a) and (b), respectively. In figure 4 (a), the outline from the hydrophone agree well for the main lobe with the pressure from optical measurement. However, less agreement is seen for the side lobes. In figure 4 (b), good overall agreement is seen.

To improve the S/N ratio at the side lobes in figure 4 (a), a longer propagation length was chosen and the result is shown in figure 5. Here, the amplitude was normalized based on the absolute peak pressure in figure 4 (a). With the long optical propagation length, the side lobes agree well, but the reconstruction seems to have failed in the main lobe because the assumption for eq.(1) started failing.

To solve this problem, the two sets of shadowgraph data with optical propagation lengths of 50 and 280 mm for the main lobe and side lobes, respectively, were combined. The reconstructed pressure on the axis is shown in figure 6. Good agreement can be seen for both main and side lobes. In this way, the pressure field up to 3 Mpa_{pp} was successfully measured.

In comparison to the hydrophone measurement, the absolute pressure from the optical measurement was about 50 %. Although absolute measurement should be possible by the proposed method in principle, calibration is still needed at this stage. In order to establish an absolute pressure measurement method, we have to solve this problem.

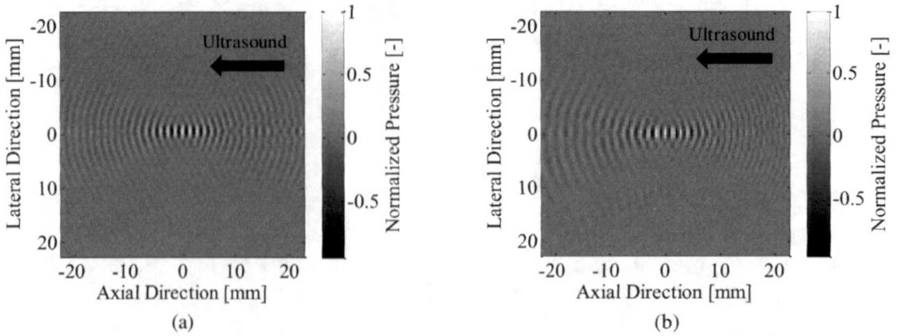

Figure 3. Normalized absolute pressure field (a) optical and (b) hydrophone measurement.

4. Conclusions

We reconstructed an ultrasound pressure field from background-subtracted shadowgraphs. Using a holographic diffuser, the optical propagation length was precisely defined. Using two sets of shadowgraphs with optimal optical propagation lengths, we successfully reconstructed an ultrasound pressure field identical to hydrophone measurement up to 3 MPa_{pp}. Some problem is still remaining, which must be solved to establish an absolute pressure measurement method.

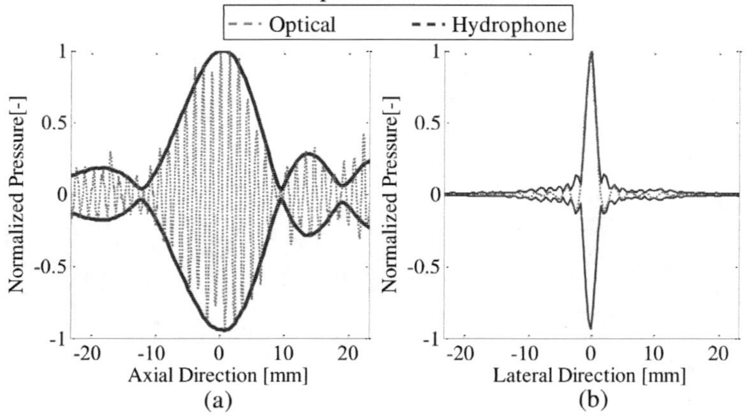

Figure 4. Comparison between two methods for *l*=50 mm. (a) axial and (b) lateral direction.

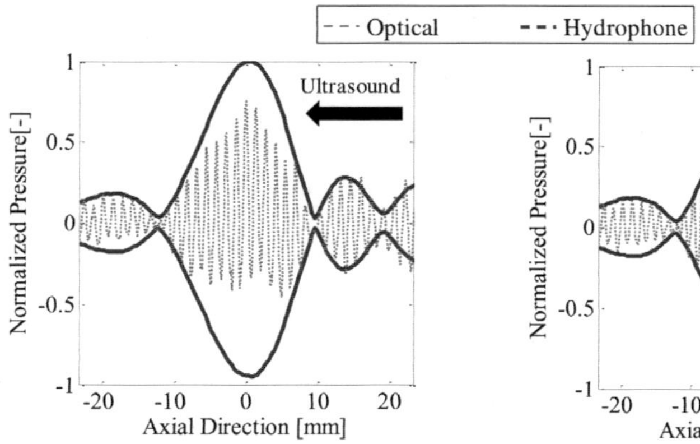

Figure 5. Comparison between two methods for *l*=280 mm.

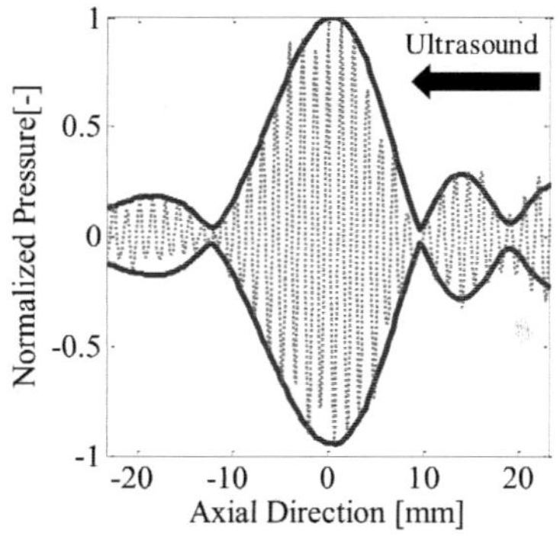

Figure 6. Comparison between two methods for the combined data.

References

[1] T. A. Pitts and J. F. Greenleaf : J. Acoust. Soc. Am. **108**(2000) 2873.

[2] H. Ouchi, N. Kudo, K. Yamamoto, and H. Sekimizu: IEICE Tech. Rep. **19**, 9 (2004). [in Japanese]

[3] R. Omura, Y. Shimazaki, S. Yoshizawa and S. Umemura: Jpn. J. Appl. Phys. **50**, 07HC07 (2011).

[4] T. A. Pitts, A Sagers, and J.F. Greenleaf: IEEE transactions on ultrasonics, ferroelectrics, and frequency control, VOL. 48, NO. 6, November 2001

[5] Erwin K. Reichel and Bernhard G. Zagar: IEEE transactions on instrumentation and measurement, VOL. 55, NO. 4, August 2006

[6] S. Harigane, R. Miyasaka, S. Yoshizawa and S. Umemura Jpn. J. Appl. Phys. **52** (2013) 07HF07

[7] M. J. Crocker: Encycl. Accoust. **1** (1997) 1

AUTHOR INDEX

Institute of Physics
Dirac House, Temple Back
Bristol BS1 6BE UK

CURRAN ASSOCIATES INC
proceedings
.com